Welcome . . .

ALSO BY CHARLES MacPHERSON

The Pocket Butler

The Butler Speaks

CHARLES MACPHERSON

THE Pocket Butler's Guide to Travel

Essential Advice for Every Traveller,
From PLANNING *and* PACKING
to MAKING THE MOST
of YOUR TRIP

appetite
by RANDOM HOUSE

Library and Archives Canada Cataloguing in Publication is available upon request.

ISBN: 978-0-147-53086-8
eBook ISBN: 978-0-147-53087-5

Book and cover design: Five Seventeen

Cover images: (travel tags) Mosquito/Shutterstock.com; (butler) © ImageZoo/Corbis

Book illustrations: Sean Kelley, courtesy of Charles MacPherson Academy Inc.

Printed and bound in China

Published in Canada by Appetite by Random House®, a division of Penguin Random House Canada Limited.

www.penguinrandomhouse.ca

10 9 8 7 6 5 4 3 2 1

appetite by RANDOM HOUSE | Penguin Random House

To my mother,

Patricia Denise Mathan.

Traveller, gourmand and connoisseur of life.

CONTENTS

Introduction

THE VERY FIRST TRIP I remember taking was to Martinique, in the French West Indies, when I was a child. My parents travelled often, and always brought me along with them. Two things about that trip stand out in my mind: first, that we got dressed up to take the airplane, and second, that the meal served on board was absolutely delicious. My mother felt the same way about airplane food back in those days, although she was adamant that the food served on flights from France was much better than on flights from anywhere else! It's so funny to think of it now—can you imagine the food being the highlight of a flight?

Airline travel has changed. My dear friend Karen Walker was a flight attendant for Pan Am in the 1960s and '70s. She loved working in the first-class cabin, and she tells stories of wheeling the trolley with a full standing rib roast directly to the seated passengers, then carving to order. It is a far cry from today's buy-on-board selections, or a miniature bag of pretzels!

My parents travelled the world extensively and always took me with them, without hesitation. My most memorable childhood trip was to the exotic destination of Tunisia in north Africa. On our first night there we went to a local restaurant (I can still picture the room). The place was packed and noisy and I, a nine-year-old, was tired. I fell asleep on the banquette beside my mother, with her coat draped over me for a blanket. My parents carried on enjoying their dinner, and nobody batted an eye at the little one fast asleep beside them.

The next day, my parents hired a driver and we spent the day visiting historic sites. I remember saying "Who cares?" as my parents dragged me from one place to another. I just didn't understand what they were so excited about. To me, one crumbling ruin looked like any other. But my parents were relentless and insisted that we could not go back to the hotel until the day was through; I just had to keep up with them. In hindsight, I'm glad I did, because I believe my passion for travel, and the art of travelling well, started way back then.

Today I consider myself a professional traveller. I spend a tremendous amount of time travelling for business: in 2016 alone I logged over 109,000 air miles! I'm not sure if this is a badge of honour or a badge of insanity, but it does mean that I have learned a lot about how to travel well. And, as a professional butler, I have a wealth of experience to share concerning how best to ease the trials and tribulations of travel.

I do hope you find my tips and notes on the following pages useful, and at least some of my stories and anecdotes amusing. It is my pleasure to share them with you.

Bon voyage!

Charles P. MacPherson

PLANS, PREPARATIONS AND PACKING

"To travel is to discover that everyone is wrong about other countries."

ALDOUS HUXLEY,

Author of *BRAVE NEW WORLD*

Planning to Travel

WHY DO WE TRAVEL? For many, myself included, it is a necessity of work. There is also leisure travel, which many of us daydream about on a daily basis. And now we're seeing the rise of "bleisure" travel, which is the perfect antidote for the busy business traveller—simply extend a business trip to include some well-earned leisure time.

No matter what your reason for travelling, I believe you will make the very most out of your trip by planning and preparing well, and well in advance.

BUSINESS

Today, it is possible to book flights to anywhere in the world at the touch of a finger. The ease and speed of travel today have facilitated an exponential increase in

DID YOU KNOW?
KLM is the world's oldest airline, established in 1919. Its first flight, between Amsterdam and London, took place on May 17th, 1920.

business travel. But business travel can be exhausting and demanding, and sometimes downright stressful. As a frequent business traveller, I have found that the more disciplined I am with my sleeping and eating routine on board and on the road, the more successful (and less stressful) the trip.

BUTLER'S TIP: BUSINESS TRAVEL

Join an airline loyalty program—these can provide many benefits for frequent flyers.

Join a hotel loyalty program—these can offer perks and paybacks for frequent stays.

Subscribe to expedited immigration entry programs such as Global Entry or Nexus for faster customs clearance.

⁂

Prepare a business travel checklist (see page 58) to ensure you have everything packed. This should include all materials you will need for your client meetings, presentations and so on. Don't forget items like printed documents and power cords for computers.

⁂

Travel light, preferably with just a carry-on, and pack clothing that is versatile and colour-coordinated.

⁂

Keep your essentials stored in your luggage at home, so you do not have to repack them for every trip. This might include a full toiletries kit and extra chargers for your electronic devices.

⁂

Establish an onboard and on-the-road eating and sleeping routine. I force myself to sleep on the airplane, reduce my food intake and never drink more than one alcoholic beverage per flight. While away, I also make sure to find time for a walk, at least 30 minutes every day, to clear my mind.

I am travelling on business and worry about my client wanting to go out drinking at night. I need a good night's sleep. How do I politely refuse? I think the best way to deal with this is head-on. Thank the client for the invitation, and either agree to join them for one drink, or politely point out that you have work to do on their account so you will have to stay behind. There's always another night!

What is the best way to organize my schedule and prepare for challenges along the way? No matter how busy you may be, before you depart it's important to write out your entire travel itinerary. Be specific, and include details on air and ground transportation, car rentals, hotel accommodations, meetings and appointments, dinner reservations and any other pertinent information. Note travel time between meetings and appointments, and any time zone changes. Include all pertinent phone numbers, addresses and contact details, so that you can quickly reach people en route. A detailed itinerary will help you figure out the best use of your time and keep you on track. And if unexpected issues arise, having this information all in one place will help you to stay calm and reorganize yourself with plenty of time.

Is it reasonable for clients to expect me to work day and night while on a business trip, or can I take a few minutes to do some personal sightseeing? If you are on a business trip and travelling on the client's dime, so to speak, it would seem fair for the client to monopolize your time. That's why I have now been to Beijing five times and still have not seen the Great Wall of China! This is a subject you might want to raise with your client before travel plans are finalized. Perhaps you could arrange a bit of "bleisure" travel by adding a day or two to the beginning or end of your trip. Just remember that any additional expenses you incur while sightseeing are your own.

PLEASURE

Leisure travel is the best kind of travel. If the office can manage without me for a week or two, I'll pick a destination abroad where it is warm and sunny, someplace quiet and free of distractions where I can totally relax. That is my perfect vacation, but it may not be yours. Think for a moment about what you really need from your next vacation; close your eyes and picture yourself somewhere free from the responsibilities of everyday life. What do you see? If your idea of the perfect holiday involves peace and quiet, you might plan to travel solo or with a like-minded friend, staying in a smaller hotel or a privately rented home. A home exchange would also give you the privacy you prefer. But if you enjoy socializing and planned activities, consider a cruise, a group tour, a large resort hotel or a sociable B&B. A travel professional could be a big help in putting together just the right kind of trip.

I have been invited to vacation with someone I truly don't want to travel with. How do I politely decline? In this type of situation, many people rely on a white lie, such as an excuse about work, a lack of funds or a previous obligation. I think being honest is easier, and that way, you never forget what you said. I personally would try something like, "Thank you so much for the invitation. A vacation in Greece next spring sounds wonderful, but to be honest, the next trip I'm planning is a solo excursion to climb Machu Picchu. It has been a dream of mine since I was a child."

My travel companion and I cannot agree on our next destination. How do we reach a compromise? Before you agree to travel with anyone, take some time to discuss your travel expectations and priorities. If your idea of heaven is skiing in the Rockies, but your travel partner wants to shop for shoes in Milan, a compromise might not be possible, and taking turns choosing holiday trips will be your best option. Remember, though, if you have different interests, you can agree to spend some time apart on the trip doing the activities you prefer. There are ski resorts just a couple of hours outside of Milan, aren't there?

HOW TO BOOK: TRAVEL AGENT OR DO IT YOURSELF?

As a business traveller, I have very strong feelings about this: I always use my travel agent, Susan Goldberg, whom I absolutely adore! As a travel industry professional, she always knows how best to guide me, even if it's just a simple day trip to New York City. And if problems arise, she is ready to step in, saving me precious time and often money. One time I was in Mumbai, India, and foolishly missed my flight; she made all the necessary arrangements to change my ticket and get me on the next plane, handling the service fees and more. She got me to my next destination on time for my client meeting! At times like these, travel agents are worth their weight in gold.

GOLDEN RULES OF TRAVELLING ALONE

No matter who you are, travelling alone can be daunting, and even scary at times. Pre-planning and researching your travel destinations will help you to navigate your new surroundings with confidence and not stand out like the typical lost tourist. Here are some tips for travelling solo:

⬩◆⬩

Plan ahead. Know as much as you can about where you are going, including which areas are safe for a solo traveller. When in doubt, consult the hotel staff.

⬩◆⬩

Know the local customs and respect cultural differences. For example, dressing inappropriately can be extremely offensive in some countries.

⬩◆⬩

Always trust your instinct. If it doesn't feel right, most likely it's not.

⬩◆⬩

Seize the opportunity. The advantage of solo travel is the freedom to do all the things that you want to do, at your own pace. No need to worry or hesitate about time spent at the museum, shopping, strolling around or just watching the world go by at a café. I'm never bothered by dining solo, either—it is a treat to eat at any restaurant I choose!

◆◆◆

Be open to meeting new people. It is often easier to meet new people when travelling alone. Do remember to consider your safety, and meet up with your new acquaintances somewhere public, like your hotel lobby. Don't venture to unfamiliar or out-of-the-way places where you might find yourself in a vulnerable situation.

◆◆◆

Make your whereabouts known. Leave a copy of your itinerary—complete with flight, accommodation and contact details—with someone at home. You should carry your hotel's business card with you at all times, as well as the contact information and addresses for any people you know locally and for your embassy or nearest consulate office.

Where to Stay

NO ONE TYPE OF ACCOMMODATION is better than another; which type you should choose really comes down to your personal preference, objectives and budget. Generally speaking, I prefer to stay in a hotel whenever possible—although of course a yacht is a wonderful option for a holiday as well! For pure relaxation in a social setting, you may consider a cruise or all-inclusive resort. If you enjoy a more authentic local experience, vacation rentals (booked through websites such as Airbnb or VRBO) or home exchanges can be a good option.

	Hotel	All-Inclusive Resort	Vacation Rental	Home Exchange
Pros	Hotels offer the most privacy and amenities, with professional management and daily housekeeping.	Resorts offer a safe, easy, cost-effective vacation for large groups, with unlimited food and drinks and a variety of activities and entertainment.	A vacation rental is usually less expensive than a hotel while still providing privacy. It also allows you to live among the locals.	The home exchange is a potentially low-cost, high-quality option that allows you to live among the locals.
Cons	The cost is high, bookings and check-in/check-out times can be rigid, and you will need to have all your meals in restaurants.	Food options are often bland, settings can be noisy, and spending time outside the resort means you incur additional costs. Provides a limited experience of the local culture.	The cleanliness and the amenities offered are often at a lower standard than other types of accommodations, and travellers must be fairly self-reliant.	Cancellation by either party causes huge problems, amenities are likely limited, and you must open your own home to strangers.
Summary	Hotels are the top choice for both business and leisure travel if you desire full amenities.	All-inclusive resorts are ideal for large groups whose main objective is relaxation and socialization.	Travellers seeking an authentic local experience at a lower price may prefer a vacation rental.	Home exchanges work well for people who want an authentic local experience and are comfortable sharing their home with others.

B&B or Guest House	Friends or Family	Private Yacht	Cruise
B&Bs offer a warm and personal experience (including home-cooked meals) at a lower cost than hotels.	Staying with friends or family is extremely low-cost and allows you to visit with those you love.	Yachts offer an extremely peaceful and luxurious stay and the chance to visit different destinations.	Cruises require very little planning and have activities to appeal to all ages.
Guests are expected to interact with each other and the hosts, and may have to share facilities.	There is very little privacy, you'll have to clean up after yourself, and you'll be subject to your host's schedule.	The cost is high, comfort is subject to the elements, seasickness may occur, and you are confined to a small space.	Seasickness may occur, you are confined to the ship while at sea, and you will likely have to dine with others.
B&Bs and guest houses are good options for moderately social travellers.	People who have friends or family living in desirable destinations can find this a very inexpensive way to travel.	Experienced sailors or those seeking a particularly luxurious voyage might especially appreciate travel by yacht.	Cruises are an excellent option for seniors or time-strapped families with children.

Here are some useful tips for tackling common challenges and questions that come up when booking accommodations.

HOTEL

My friend wants to share a hotel room to save money, but I don't want to do this. How do I tell her? Hotel accommodations are generally expensive, and many people propose sharing a room to save money. However, if you absolutely don't want to do this, then don't skirt around the issue. You can say something like, "You know I am a creature of habit, and I hope you don't mind, but I'd be much happier staying in my own room."

How do I get the best from my hotel budget? If I have a choice between the cheapest room in a five-star hotel or the Presidential Suite in a two-star hotel, I'll always pick the former. Even if the room is smaller, you will still experience the premium service and amenities that are offered only by a five-star hotel. This is a great compromise for me!

ALL-INCLUSIVE RESORTS

Are there different kinds of all-inclusive resorts, or are they all the same? There are many different kinds of all-inclusive resorts. Some cater to adults only, some couples only, some families and some singles. Do your research, read reviews and call the resort if you still have questions.

Am I stuck at an all-inclusive or can I leave the resort? All-inclusive resorts generally prefer you to stay on the premises to enjoy their activities, or purchase their tours and other special amenities. Having said that, renting a car and doing your own thing is absolutely an option. Just remember that while you are on the road you may be incurring expenses for snacks and meals that you have already technically paid for at your resort. In fact, if you plan to venture out and be on the road exploring every day, the all-inclusive resort may not be an ideal choice for you.

VACATION RENTALS

The website listing offers a lot of information, but how can I be sure that the vacation rental I choose will be right for me? When I have chosen this option in the past, I have always taken the time to contact the potential host in advance of booking, either by email or by telephone. This allows me to ask specific questions and find out if the accommodation and the neighbourhood offer everything I need.

Can I pay by cash to save money? Vacation rental companies protect you with a variety of rules and regulations, but they only apply if you pay through their site. Paying cash or otherwise attempting to circumvent the rental company is a bad idea and can put you at risk.

What if I'm not happy? Who do I speak to? Always communicate the issue to your host first. The host may be able to solve it for you quickly. If the host is unable to resolve the issue, contact the website you booked through directly.

HOME EXCHANGES

How do I know if I am making a good exchange? This is hard because the short answer is you don't. Using an online agency with personal reviews helps. Speak to the potential host on the phone ahead of time to get a sense of what you might expect.

What if I damage something or, worse, what if the other party damages something in my home? Prior to the exchange you need to discuss and establish a policy for breakage and damage. Generally speaking, the person who did the damage should be held responsible for replacement or repair. That said, I think it is very important that anything of value (including, especially, sentimental value) needs to be put away. Additionally, make sure you have proper coverage through your home insurance.

How do I deal with personal items in my home? People will snoop in your home. You need to accept this from the start and prepare accordingly. Lock away or remove anything you want to keep private, and be sure to protect yourself from identity theft by removing sensitive information like bank and credit card statements, utility bills and documents related to investments, and putting a hold on incoming mail.

B&B/GUEST HOUSE

Is it rude to take breakfast to my room? This depends on the establishment, but it is generally frowned upon. You are usually expected to eat at the table because interacting with the host and other guests is part of the experience of staying at a B&B.

Can I go to bed when I want? Yes, but you must be respectful of others. If you're a night owl, check before booking that the establishment does not impose a curfew, and, when you go to bed, be mindful of others who may already be asleep.

FRIENDS AND FAMILY

Do I have to eat what is being served? Well, let's say you absolutely must try! Your host has gone to the trouble of preparing home-cooked meals for you, so do try to be gracious. Be sure to communicate any dietary preferences before you arrive.

How do I deal with my host family members and paying for expenses? Money can be a touchy subject, and I think it is best to raise this issue prior to your stay to avoid misunderstandings and hurt feelings. Always offer right away to pay for all your expenses. If your host is reluctant to take money from you, you may wish to offer to cook a meal or take your host out to dinner as a thank-you.

I'll be staying at the home of friends while they're out of town. Do you have any tips on being a good guest for my absentee hosts? Don't be shy about asking for special instructions before you arrive. Do not help yourself to the liquor cabinet (or anything else, for that matter), or if you do, be sure to replace (exactly) what you use. Be considerate and keep the place tidy; strip the bed and launder the bedding before you leave. At the end of your stay, leave a thank-you note and a small gift.

PRIVATE YACHTS

Who should I speak to if there is a problem on board? The captain is the ultimate authority on board, not the chief steward. If there is a problem, speak to the captain.

What are the expectations for tipping on a yacht? The crew works very hard to please, so the tip should reflect their performance. The rate is generally 15 to 20 per cent of the charter fee. It is customary and preferable to give the captain the gratuity in cash in an envelope at the end of your voyage, and he or she will distribute it to the crew.

CRUISE SHIPS

Is it a good idea to take kids on a cruise? That depends. Just like resorts, different cruise ships cater to different customers—parents with kids, adults only, seniors. So, when you do your research, keep this in mind.

What do you do all day on a cruise? Well, it depends on the cruise line you've selected, so this is something to research. Provided you picked the one that best suits your interests, there will be plenty of appropriate activities to entertain you, and the cruise director will always be happy to help you find something to do.

DID YOU KNOW?

Cruise ships use their own emergency codes to alert the crew, including: "Code Blue" for a medical emergency; "Bravo, Bravo, Bravo" for a fire on board; "Mr. Mob" or "Oscar, Oscar, Oscar" for a man overboard.

GOLDEN RULES OF TRAVELLING WITH THE ELDERLY

Dear friends who have travelled with their elderly parents have told me how fulfilling it is, but also jokingly remarked how similar it can be to travelling with small children. Trips are slowed down to a pace that the oldest companions can manage. You may find yourself attending to their most basic needs, from snacks to naps, and keeping a watchful eye as they make their way to the loo. It takes patience, but it is worth it in the end! Here are some tips my friends have shared with me:

◆◆◆

Pick senior-friendly destinations that are easy to navigate. If your elderly companions walk with a cane or walker, the smoother the walking surfaces the better. You'll need to save those lovely cobblestoned streets for another trip.

◆◆◆

Travel insurance with medical coverage is a must. Have a list of all medications and drug allergies, and be familiar with your elderly companions' medical conditions and histories. Be sure to have contact information for their doctors handy.

◆ ◆ ◆

For those who rely on wheelchairs or other mobility aids, contact the airline for specific details on transport arrangements. Air travel requires a substantial amount of walking, standing and waiting, so even those who usually walk unaided may find it helpful to have the assistance of a wheelchair.

◆ ◆ ◆

Travel during the day. Very early mornings and evenings can be taxing for the elderly, so travel when their energy levels are highest.

◆ ◆ ◆

Arrange for a car service. Plan ahead for comfortable and convenient transport between the airport and your accommodations.

Don't be too ambitious with the travel itinerary. Limit your day to one or two activities at most. The elderly may tire easily, so make sure to plan rest breaks, snack breaks or nap times. Setting aside an entire day in the itinerary for rest can also be helpful.

DID YOU KNOW?

On the luxury cruise liner Queen Mary 2, a list of 100 things to do the following day is left on your pillow every night.

Preparing to Travel

THE BOY SCOUT MOTTO, "Be Prepared," applies to all facets of life. I can tell you from experience that there is nothing that helps ensure a trip's success more than thorough preparation. And knowing that I can meet the inevitable challenges of travel head-on gives me peace of mind.

TRAVEL TIMELINE

I'm a list person, so I feel very calm and in control when I have a checklist and I can tick off items as they get done. On the next pages I have provided a good basic travel-prep timeline for you to start with. Feel free to make any adjustments you need to fit your situation.

TWO MONTHS BEFORE DEPARTURE

- If you have any outstanding or ongoing health issues, schedule a check-up with your physician and dentist.
- When planning travel to foreign destinations, check with your doctor or a travel clinic and get any recommended vaccinations.
- Review your travel documents to make sure they are in order; check that expiry dates are a minimum of six months out from your return date.
- Review visa requirements and make the necessary applications.

ONE MONTH BEFORE DEPARTURE

- If you're travelling out of the country, advise your bank or credit card company of your travel plans.
- Make sure you have the correct luggage to meet the needs of your trip (see page 48); measure your carry-on to ensure it meets the airline's standard.

TWO WEEKS BEFORE DEPARTURE

- Review your itinerary to ensure everything is booked; print out two copies—one for yourself and another to leave with someone at home in case of an emergency.
- Start to think about your clothing needs based on your itinerary.
- Make sure you have adequate supplies of any prescribed drugs, and if not, renew or refill your prescriptions.

SEVEN DAYS BEFORE DEPARTURE

- Review your toiletries and determine what needs to be replenished so that your travel kit is fully stocked.
- Check that you have the necessary power cords for all your electronic devices, and if travelling abroad, any power adapters you might need for your destinations (see page 56).
- If you will need a cellphone plan while travelling, call your provider to arrange the best option for your trip.
- Pull out the luggage and open it to air it out.
- Remove any old baggage tags and airline stickers.

- Make sure your luggage tags are intact, and check to make sure you have your name, address and cellphone number on the inside of the suitcase. Should your luggage tag get lost, this will enable the airline to track you down.
- Order foreign currency, or pick up a prepaid currency card. Pay down or pay off your credit card balance so that you have lots of spending room available.

SIX DAYS BEFORE DEPARTURE

- Gather your travel clothes, shoes and accessories to review what you intend to pack.
- Do any required laundry or dry cleaning of items to be packed for the trip.

FIVE DAYS BEFORE DEPARTURE

- Prepare a detailed list of all the items you're planning to pack, including accessories and shoes.
- Put a hold on your newspaper delivery, if applicable.

THREE TO FOUR DAYS BEFORE DEPARTURE

- Do any last-minute errands, including any beauty treatments. This is a good time for a haircut, manicure or pedicure.
- Pack your suitcases, checking everything against your packing list.
- Check for travel advisories.

ONE TO TWO DAYS BEFORE DEPARTURE

- Check the weather forecast for your destination.
- Pack toiletries into your suitcase.
- Pack any liquids for the carry-on as per the 3-1-1 rule (see page 57).
- Pack your carry-on bag with your wallet, passport, tickets, reading materials, cellphone, charger cables and any other in-flight essentials.
- Check flight details and print out boarding passes.

TRAVEL ESSENTIALS

There is nothing worse than being unprepared or caught by surprise when travelling, especially in a foreign land. Be sure to keep these items with you whenever travelling (see checklist page 58).

PASSPORT

It is always a best practice to keep your passport up to date. Different countries have different rules, but as a general guideline you should make sure that you have a minimum of six months left on your passport before expiry. The six-month period should be based on your return date, not the departure date for your trip. It is also important to remember that some countries require at least two full blank pages in your passport. I put a reminder in my calendar to renew my passport when it's no more than nine months from its expiry date so I'm not scrambling at the last minute.

VISAS

Having a passport does not guarantee you entry into every country. A passport is an official document issued by your country that proves your identity for the purposes of travel, while a visa is an official document issued by a foreign government allowing a traveller to enter and temporarily stay in that country. I have been caught a few times scrambling for a travel visa for last-minute business trips—it's always stressful.

When it comes to obtaining travel visas before your trip, there are two options: you can go to the embassy or consulate of the country you are travelling to and apply for the visa yourself, or you can engage a company that offers visa consulting and processing services. For a fee, they will assist you with forms and any other requirements and obtain the visa on your behalf. This is a very practical and convenient service for business travellers and people who are pressed for time.

MEDICAL PREPARATIONS

Bring a full list of all your medications and medical issues, as well as proof of all relevant vaccinations, and contact information for your medical practitioners. This might all be helpful in the event of an illness abroad. If you have a pre-existing medical condition, bring a list of the medical facilities at your destination that would be appropriate for any issues that might arise.

TRAVEL INSURANCE

Often, I am asked, "Do I need travel insurance?" I think, for peace of mind, the answer is always yes. Your credit card may cover some basic expenses; however, is it enough? Good travel insurance coverage can include trip cancellation, emergency medical, reimbursement for lost luggage, rental car collision and more. To me, having emergency medical coverage is always a must. Being sick away from home is never pleasant, and medical treatment in a foreign country can be very expensive. Even a basic policy is an excellent investment, in my opinion. Be very clear before you leave, however, about what it does and does not cover, especially when it comes to pre-existing conditions.

Two important points on travel insurance:

1. Hospitals and clinics in some countries will refuse to treat foreign patients if they do not have adequate travel health insurance or money to pay the bills.

2. Travel advisories can affect your insurance coverage policy before you leave, or while travelling abroad. Read the fine print to understand exclusion clauses that may invalidate your policy if an official travel advisory is issued.

EMBASSY AND OTHER EMERGENCY INFORMATION

If you are going to a dangerous part of the world, always register with your government so that, in the event of a crisis, they know where you are. This is often referred to as the Registration Abroad Service. And, regardless of your destination, it is always a good idea to keep a note with the address and phone number of the closest embassy or consulate of your home country, in case of emergency.

You should also have a list of emergency telephone numbers in case you lose your phone or it stops working. The list should include contact details of your booked accommodations and local transportation, emergency numbers for credit and debit cards, information about your travel insurance, and names and numbers for medical doctors and family members.

MONEY

There was a time when the American Express Travelers Cheque was the currency of choice for travelling, and I remember those days like they were yesterday. Now, however, there are far more convenient and secure options.

I believe that you should never restrict yourself to just one option when it comes to money. For one thing, credit and debit cards, though very useful, are still not accepted everywhere in the world. And having the local currency is always handy for when you need to pay for meals, taxis, tips and especially the luggage trolley upon your arrival.

Method of Payment	Pros	Cons
Cash	Convenient. Best in emergencies. Always accepted.	Theft: once it's gone, it's gone. Where is it safe to keep it?
Debit Card	Accepted at most bank machines. You can only spend what you have in the bank. If lost, your money is reasonably safe.	Bank fees. Limited number of places where they are accepted. Possible card failure. Purchases not protected.

Method of Payment	Pros	Cons
Debit Card (continued)	No interest or late fees charged. You have access to your transactions online.	No access to extra funds. You often need to notify bank of travel plans.
Credit Card	Accepted at most shops, banks and bank machines. If lost, your money is reasonably safe. You have access to extra funds, over and above what is in your bank account. Cash-back options. You have access to your transactions online.	Can overspend if not careful! Bank fees. Exchange rates are not always favourable. Possible card failure. You often need to notify bank of travel plans.
Prepaid Currency Card	Convenient. Access to your bank to transfer funds as needed. If lost, your money is reasonably safe. Pre-load with different currencies.	Bank fees. Exchange rates are not always favourable. Possible card failure.
Traveller's Cheques	Safe. Will be refunded if lost or stolen. Can be purchased in different currencies.	Not as widely accepted as in the past. High fees.

BUTLER'S TIP: TRAVEL MONEY

Purchase foreign currency before your trip to get the best exchange rates.

◆ ◆ ◆

Inform your bank or credit card issuer that you'll be travelling so they will be aware of your transaction activity. You don't want to have your card blocked while you're away.

◆ ◆ ◆

Make sure to bring with you a list of credit and debit card contact numbers in case of emergency. For traveller's cheques, bring with you a record of the serial numbers in case they are lost or stolen.

◆ ◆ ◆

Do not keep all your money in one place.
If you do, you could lose everything!

◆ ◆ ◆

Always carry some cash on your person in case of card loss, malfunction or absence of bank machines. If possible, stash it separately in concealed pockets of your clothing.

DRIVER'S LICENCE

Make sure your driver's licence is up to date prior to departure, especially if you plan on using a rental car during your trip. To drive in some countries, you may be required to have an International Driving Permit. This can also be useful as an additional piece of ID.

Being Away From Home

WHEN I AM PREPARING for a trip, I am an absolute believer in either a written to-do list (I love using Post-It notes) or a master list on my mobile phone. It doesn't matter what method you use, just don't try to keep everything in your head—you will be apt to forget something, and then you'll be stressed about trying to remember what you forgot. As a butler, I find that the most common-sense method is generally the most effective.

PLANT CARE

For the avid gardener, plant care is a serious concern. For short-term absences there are some great gadgets available, such as watering bulbs or globes for potted plants. For a garden, unless it's drought-tolerant, you will need to have watering systems in place. If you have someone who is able to tend to your high-maintenance plants or garden in your absence, be sure to leave explicit instructions.

PET CARE

Prepare a plan for your stay-at-home pets well before your date of departure. A reliable pet-sitter or a willing friend or family member is the ideal solution. Before you leave, prepare a note for your pet-sitter that includes the contact information of your veterinarian and any special health issues or dietary needs your pet has. Boarding your pet is another option, but you will want to check references and inspect the facility to feel confident that Fido or Fluffy will be comfortable.

CHILD CARE

When I worked in private service, there were several occasions when my employers had to travel without their children. There was a great deal of separation anxiety, especially when the children were young, but they were fortunate to have doting grandparents who welcomed the chance to look after them. Here are some steps to make the separation as stress-free as possible for everyone involved:

- Put together a file with contact details for the children's doctor (along with any pertinent medical information) and their school, and include other important phone numbers and addresses, possibly including a helpful neighbour who might step in in a pinch.
- Prepare a calendar with your travel dates so the children can keep tabs on your whereabouts. A map might be fun as well, if you have more than one destination.
- Prepare a calendar for the children with their own scheduled activities.
- Provide an outline of each child's daily routine and habits, which also includes expectations and boundaries.
- Check in with the children regularly. It's so much easier these days with all our electronic devices.

HOUSE-SITTERS

If you are going away for a particularly long period, consider a house-sitter: someone who can receive your mail, water your plants, mow the lawn and take care of other everyday household tasks. This could be a family member, friend or professional, live-in or live-out. Prepare detailed checklists and upkeep instructions, a list of emergency contacts and a list of your preferred service providers. If you are hiring someone, make sure to check references, do a criminal background check and have a written contract that protects both parties.

REDIRECTING MAIL

While doing research for this book, I came across the term "Poste Restante," or, as it is more commonly known in North America, "General Delivery." This is a service whereby you can have your mail redirected to the local post office branch at your travel destination almost anywhere in the world. This may be a good option to explore if you are planning to be away from home for a long period of time. Specific guidelines differ by country so it's always best to check the official postal service websites. A good option for shorter trips is to have your mail held for you for a specific duration of time, and then delivered all together on your return. Again, check with your local postal service for options and rates for this service.

SETTING EXPECTATIONS AT WORK

I am a strong believer in unplugging yourself from the office when you are on vacation. A good work-life balance is important for everyone, but I know it is easier said than done! Whether you are away from the office on holiday, or travelling for work and needing some distance from day-to-day office distractions and demands, I have found two methods that work well for me:

1. Appoint someone in your office whom you trust to take charge of things in your absence. Give them the ability to make decisions on your behalf, and ask them to call you only when a crisis occurs.
2. If you need to be connected more than the above allows, minimize your work time by calling in once a day (usually the morning) and offering this as the

opportunity to discuss any issues at hand. This allows you to call in at your own convenience. Once you've checked in, you are free for the rest of the day without having to think about the office!

When you are crafting your email out-of-office reply, write a meaningful message with politely worded, clear instructions for the recipient. Whenever I am travelling, I write a response that includes telephone numbers and email addresses of my staff, and whom to contact during my absence for different types of inquiries. The same goes for your voicemail message. This unambiguous approach will set accurate expectations about your availability and realistic response time.

Packing

WHETHER YOU'RE A minimalist packer or a take-everything packer, what and how you pack will make a big difference in your travel experience. Packing wisely will simplify your life on the road, so take time before you leave to get everything in place.

LUGGAGE

The great question pondered by warrior travellers and novices alike: Should I check my bags or carry them on? In the movie *Up in the Air*, George Clooney's character is a frequent-flyer businessman who navigates every airport in record time by never, ever checking his bag.

Advantages to this strategy include no time spent waiting at check-in or at the baggage claim, and no chance of lost luggage. If you can pack light with a carry-on, it's a good way to travel.

However, I love to check my luggage. Why? Once my bags are checked, I can relax and enjoy the trip without having to drag my luggage through the airport. And I do so much air travel, I don't need the additional stress of a fight for overhead bin space.

Ultimately, the type and length of your trip dictates the luggage you should bring.

CHOOSING A SUITCASE

Given the plethora of suitcase choices for both checked and carry-on baggage, finding the right one for you is not always easy. Here are a few things to consider.

SIZE: Can you easily move around with it? Test it out by walking through the store with the suitcase.

COST: I travel at least once a week, and so I have purchased a sturdy suitcase that is almost indestructible.

It was expensive but an investment. If I travelled less, a more budget-friendly suitcase would meet my needs.

WEIGHT: I never used to think about this, but now that weight restrictions have become strict the weight of your luggage itself is very important. The airlines don't budge on weight allowances, and, I admit, I have been caught having to remove things from my bag at the check-in counter to get my luggage down to the regulated weight.

COLOUR/STYLE: My luggage tends to be black, so it looks like everyone else's. I had an employer once who always bought the least popular colour sold in the store so that his luggage would be easy to spot. For black suitcases, attach a brightly coloured luggage strap or ribbon to help you identify it.

DURABILITY: Check the wheels, handle and zippers. Hard shell is a more durable than soft shell, and its surface is easier to clean and resistant to dirt, stains and odours.

TYPE (HARD OR SOFT SHELL): I have both soft- and hard-shell luggage, and I never had a strong preference either way until a very good friend went to Antigua for New Year's Eve. Her plane landed in St. Maarten, and while she and the others waited for their connecting flight, their luggage sitting on the tarmac, there was a downpour of rain. When she eventually arrived in Antigua she found all the clothes that she had packed in her soft-shell suitcase were drenched and damaged, while a friend who used hard-shell luggage had no issues and wore her New Year's Eve outfit as planned.

Luggage Type	Pros	Cons
Soft Shell	Easier to fit into tighter or awkward spaces. More flexible when packing souvenirs and other extra items. Tends to be lighter in weight.	Less durable. Not water-resistant. Offers less protection for fragile items. Packed garments are more likely to get wrinkled.
Hard Shell	Much more durable. Water-resistant. Packed garments less likely to arrive wrinkled. Better protection for fragile items.	Rigid construction makes it difficult to fit into awkward spaces such as overhead bins. Not as flexible when packing souvenirs and other extra items. Tend to be heavier in weight.

SPECIAL BAGGAGE ITEMS

GARMENT BAGS

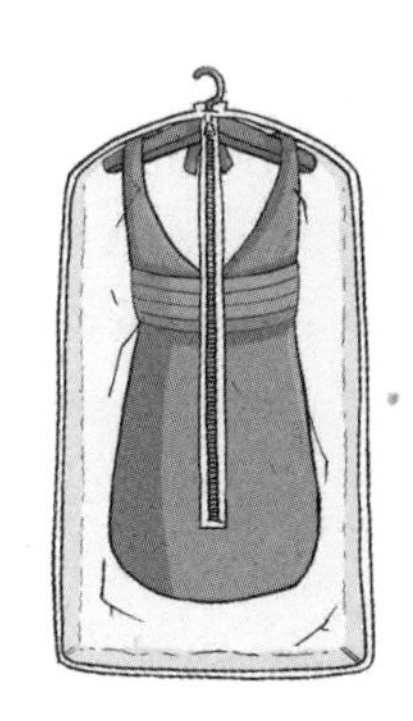

For wrinkle-free travel, garment bags are the ideal option for suits, wedding and evening dresses and any other delicate, hanging garments. Garment bags have evolved from something very basic to wheeled, foldable bags. Most bags can be brought on board as carry-on luggage, but always check size and weight restrictions with the airline first.

PET CARRIERS

Before you buy that special travel carrier for Fluffy or Fido, check pet carrier requirements and restrictions with the airline. Here are some basic issues to consider:

- Size and weight restrictions (it will count as one standard carry-on item).

- Leak-proofing and ventilation.
- Space for your pet to stand and move around safely.

OVERSIZED OR BULKY ITEMS

Will your trip require golf bags, ski bags, bicycles, camping gear, other sports and hunting equipment or large musical instruments? Most airlines will be happy to accept your oversized items, as long as you comply with their guidelines. Check with the airline for specifics on packing instructions, damage waivers and fees.

CARRY-ON BAGS

With carry-on bags, there is no one standard size so it's important to check size and weight restrictions with the airline. I have two guidelines when I'm shopping for a carry-on bag:

1. It must be small enough to fit under the seat.
2. It must have either a shoulder strap for easy carrying, or wheels.

In addition, most airlines will allow you to bring a "personal item" such as a purse, a small backpack or a camera bag. This allows you to keep items you'll require during the flight close at hand while your suitcase is stored in the overhead bin.

WHAT TO PACK

Before you begin packing, review your itinerary and determine your clothing requirements. I start preparing a packing list at this point. It's an extra step, but by being meticulous from the get-go, you'll be more likely to arrive at your destination with everything you need.

Because I'm on the road so often, I keep many of my essential travel items on hand and ready to go, so all I need to do is put them in my suitcase. As a rule of thumb, when something runs out I replace it right away. I picked up this valuable habit while working as a butler. My employer travelled every week, and it made life simpler to keep his travel toiletry kit stocked and ready to go at a moment's notice. When he returned, we would check the quantities and refill the bottles as needed. I promise you, this pre-preparation trick always helped streamline the packing process.

ESSENTIAL CARRY-ON ITEMS

When I'm travelling it's all about ease, comfort and being equipped to persevere through any travel disruption. As long as I have the following items in my carry-on I can withstand most transport delays and other headaches.

TOILETRY KIT. The absolute essentials: medication, toothbrush, toothpaste and dental floss.

CHARGERS AND CABLES. If I'll be working on the flight, I bring an electronics pouch with the chargers and cables for my devices. I always purchase a second power adapter or cable when I buy a new device, just for travelling. I also carry a backup battery charger in my checked luggage.

PLUG ADAPTERS. Often the better hotels will have plug adapters available for their guests, but if you travel abroad a lot, especially to places that are off the beaten path, consider adding a multi-plug adapter to your electronics pouch. You won't regret it! There is nothing worse than finding out that your plug doesn't fit the outlet (see page 56) just as your tablet or laptop runs out of power.

READING MATERIALS. I used to travel with at least one book, but now I just take my e-reader. I love the convenience of having access to my entire digital library, as I like to have my nose in several books at a time. Also, my e-reader gives me access to newspapers and periodicals, so I feel very connected to what is happening back home, or anywhere else in the world.

TRAVEL DOCUMENTS AND CURRENCY. I always make it a habit to carry my travel documents and currency in a separate pouch within my carry-on. This pouch holds all the essential items listed on page 58.

International Electrical Outlets

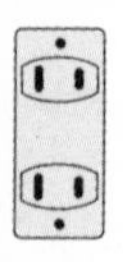 TYPE A: North America, Central America, Japan	TYPE B: North America, Central America, China, Japan	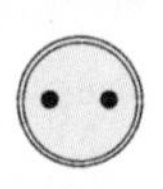 TYPE C: Europe (except the UK, Ireland, Cyprus and Malta), South Korea	 TYPE D: India, Namibia, Nepal, Sri Lanka	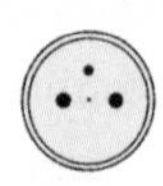 TYPE E: Belgium, France, Slovakia, Tunisia
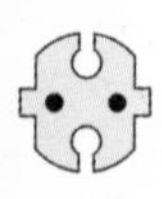 TYPE F: Austria, Germany, the Netherlands, Spain	 TYPE G: Cyprus, Hong Kong, Ireland, Malaysia, Malta, Singapore, UK	 TYPE H: Israel	 TYPE I: Argentina, Australia, New Zealand, Papua New Guinea	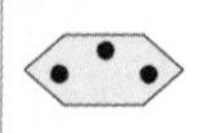 TYPE J: Liechtenstein, Switzerland
 TYPE K: Denmark, Greenland	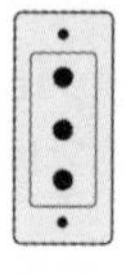 TYPE L: Italy	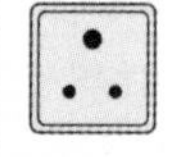 TYPE M: Lesotho, South Africa, Swaziland	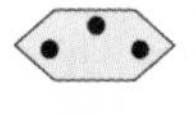 TYPE N: Brazil	

MISCELLANEOUS. There are a few other little extras (like glasses, a pen and a travel pillow) that can be beneficial to have on hand, see page 58.

3-1-1 LIQUIDS CARRY-ON RULE

The amount of liquid that can be taken on board an airplane is restricted due to the threat of liquid explosives. The global rule is 3-1-1, which means that liquids or gels must be in containers no larger than 3.4 ounces (100 ml) (that's the 3), which have to fit into a 1-quart clear plastic re-sealable bag (1), and each passenger is allowed to pack one of these (1) into carry-on luggage. Exemptions include medications, infant food and drink, and duty-free liquids. Refer to the Transportation Security Administration (TSA) website for further information.

INDISPENSABLE PACKING CHECKLISTS

With a list in hand I don't have to obsess about forgetting to pack my socks or other small but essential items. Travel checklists are especially useful since I save them for future reference. They are truly indispensable!

TRAVEL DOCUMENTS/CURRENCY

- ❑ Passport
- ❑ Visa
- ❑ Travel and medical insurance
- ❑ Embassy information
- ❑ Driver's licence
- ❑ Credit card/debit card
- ❑ Cash

CARRY-ON ESSENTIALS

- ❑ Prescription medications (properly labelled)
- ❑ Written list of emergency telephone numbers (page 39)
- ❑ Glasses
- ❑ Cellphone
- ❑ Reading material or device
- ❑ Chargers for all devices
- ❑ Travel plug adapters
- ❑ Laptop
- ❑ Headphones
- ❑ A pen
- ❑ Water (purchase post security clearance)
- ❑ Travel pillow
- ❑ Eyeshade, earplugs
- ❑ Tissues
- ❑ Valuables, such as jewellery
- ❑ Gum or hard candy
- ❑ An energy bar or snack

TOILETRIES—ESSENTIALS

- ❑ Toothbrush
- ❑ Toothpaste
- ❑ Dental floss
- ❑ Deodorant
- ❑ Soap
- ❑ Shampoo, conditioner
- ❑ Moisturizer
- ❑ Brush and comb
- ❑ Hair-styling products
- ❑ Shaving supplies
- ❑ Makeup, makeup remover
- ❑ Feminine hygiene products
- ❑ Birth control
- ❑ Nail file, nail clippers (checked luggage only)
- ❑ Cologne/perfume
- ❑ Pain reliever

TOILETRIES—OPTIONAL

- ❑ Contact lenses
- ❑ Saline solution
- ❑ Hand sanitizer
- ❑ Tweezers
- ❑ Mouthwash
- ❑ Lip balm
- ❑ Vitamins
- ❑ First-aid ointment
- ❑ Bandages
- ❑ Sunscreen

CLOTHING—BASICS

- ❑ Underwear
- ❑ Socks/stockings
- ❑ T-shirts
- ❑ Bras
- ❑ Sleepwear
- ❑ Dress shirts
- ❑ Casual shirts
- ❑ Casual pants/jeans
- ❑ Sweatshirts
- ❑ Dresses

- ❑ Skirts
- ❑ Blouses
- ❑ Sweaters
- ❑ Jacket/coat
- ❑ Hat
- ❑ Gloves
- ❑ Scarves or shawls
- ❑ Walking/casual shoes
- ❑ Dress shoes
- ❑ Belts
- ❑ Ties
- ❑ Jewellery
- ❑ Purses
- ❑ Accessories
- ❑ Umbrella—collapsible

WORKOUT GEAR

- ❑ Running shoes
- ❑ T-shirts
- ❑ Shorts
- ❑ Jogging pants
- ❑ Sport socks
- ❑ Outerwear for cool climates if exercising outdoors
- ❑ Cap or visor
- ❑ Headband, wristbands

POOL, BEACH, CRUISE

- ❑ Beach bag
- ❑ Sunscreen/sunblock
- ❑ Zinc oxide
- ❑ Lip balm with sunscreen
- ❑ Insect repellent
- ❑ Hat
- ❑ Sunglasses
- ❑ Beach towel
- ❑ Bathing suits
- ❑ Flip-flops/sandals
- ❑ T-shirt or cover-up clothing
- ❑ Shorts
- ❑ Water bottle
- ❑ Insulated bag for snacks

PACKING TO DRESS CODE

If you are travelling for business, or for a specific or formal event, it is important to know the dress code so that you can pack appropriately. If you have the choice, it is always better to be overdressed than underdressed. Picture this scenario: you're at a business casual event and you're the only gentleman wearing a tie. What do you do? Discreetly remove your tie, put it in your pocket and blend in. However, if you arrived dressed too casually, and find all the other men are wearing ties, you will definitely be the odd man out. Below is an overview of what each dress code truly means.

WHITE TIE—MEN (FRENCH: *TENUE DE SOIRÉE*; ALSO KNOWN AS "FULL EVENING DRESS")

- Black double-breasted tailcoat with silk peak lapels, worn open
- Black tailcoat with two buttons at the back, a remnant of a time when the tails were folded up and buttoned for ease of horseback riding

- Matching pleated trousers with single strip of black satin or braid in America, or two stripes in Europe
- Black suspenders for trousers
- White cotton piqué wing-collared shirt with stiff front
- Low-cut white cotton piqué vest
- White cotton piqué bow tie
- White cufflinks and shirt studs
- White or grey gloves
- Black patent leather opera slippers, black dress socks

BLACK TIE—MEN (FRENCH: *CRAVATE NOIRE*; ALSO KNOWN AS "EVENING WEAR")

- Black tuxedo or dinner jacket with satin peak lapels
- Matching black flat-front trousers
- White French-cuffed formal shirt or pleated tuxedo shirt
- Black bow tie
- Black cummerbund or black vest
- Black patent leather shoes and black dress socks

	Men	Women
Black Tie Optional	Tuxedo is appropriate; however, a dark suit, with a crisp white dress shirt and dark tie, is acceptable.	Formal evening wear, dress, cocktail dress or dressy separates.
Semi-Formal	A dark suit, with a crisp white dress shirt and dark tie.	Short afternoon or cocktail dress, or long dressy skirt and blouse.
Business Attire (aka Tenue de Ville or Lounge Suit)	Suit, dress shirt and tie.	Business suit.
Business Casual	Slacks with an appropriate sports jacket and an open-collared shirt. Business casual is classic, neat and professional; err on the conservative side.	Reasonable-length day dress, pantsuit or neat and pulled-together separates such as slacks or skirt with an appropriate blouse.

- White silk or linen handkerchief in breast pocket (optional)
- White dinner jacket with black trousers is acceptable in tropical locations, on a cruise and perhaps in the summer
- Black cufflinks and shirt studs

WHITE TIE—WOMEN (FRENCH: *ROBE LONGUE*)

- Formal full-length evening gown
- Shawl or elbow-length opera gloves if gown is sleeveless
- Jewels

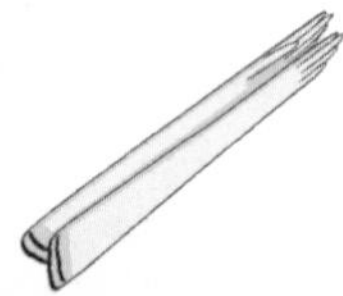

BLACK TIE—WOMEN

- Formal evening gown or very dressy cocktail dress
- Jewels

HOW TO PACK

Everybody has their own method of packing. Personally, I find the garment bag (page 51) is the best wrinkle-free option for travelling with suits, cocktail dresses and other more formal or delicate garments. When packing a suitcase, though, I combine a few different methods, including folding, layering, rolling and packing cubes.

PACKING METHODS

FOLDING. Clothes are folded flat, neatly stacked and placed in the suitcase. For convenience, stack them in the order you'll be wearing them, so that your first day's garment is on top. This is a compact way to pack.

LAYERING. This is ideal for business travellers who want as few wrinkles as possible in their packed clothes. Garments can be laid out flat and placed one on top of each other. Use tissue or plastic dry cleaning bags between the layers for extra padding and to reduce folds, creases and wrinkles. Don't worry about any garment and tissue overhang as they will be folded in at the end.

BUTLER'S TIP: TRAVELLING LIGHT

Review your packing list and lay out everything you intend to bring before you start to fold and pack. This will help you to edit out the non-essentials.

Pack mix-and-match pieces in a monochromatic colour scheme so that your wardrobe is interchangeable.

Plan on doing a little laundry on the road. This will allow you to pack less.

ROLLING. Save this method for casual clothing when wrinkles aren't an issue. Basically, folded clothes are rolled up tight, then placed side by side in the suitcase. It isn't the most space-saving means of packing, but it works well for backpacks and duffel bags.

PACKING CUBES. These are zippered, soft-sided fabric containers. They are great travel organizers and come in different sizes and shapes. You can sort your clothes and accessories by cube so that unpacking or finding that particular shirt becomes straightforward and easy. I recently bought these and have fallen in love with them!

FOLDING METHODS

SKIRT

Lay the garment face down on a flat work surface.

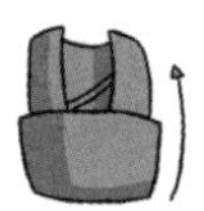

Fold each side of the skirt in ¼ of the way. Fold up the bottom half of the skirt until you reach the waist. The skirt is now folded in half.

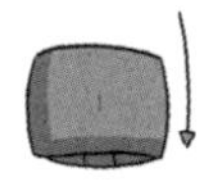

Flip over and lay flat in the suitcase.

PANTS (ROLL METHOD)

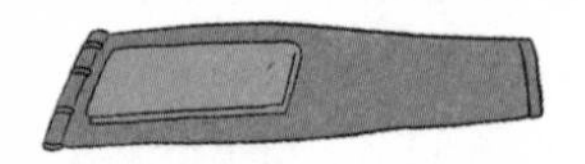

Lay the pants on a flat work surface and put a piece of tissue paper on the top half of the garment.

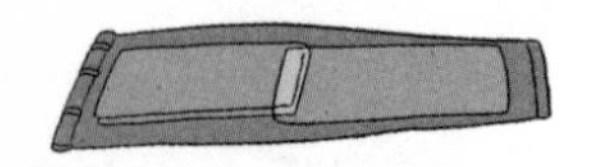

Place another piece of tissue paper so that you have tissue covering the entire length of the pants.

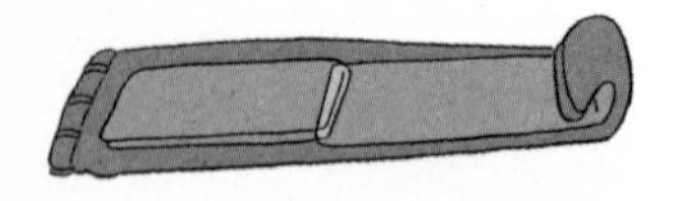

From the end of the pant leg, begin rolling the pants very loosely. Remember, a tight roll will create wrinkles.

Continue rolling loosely to the top of the pants.

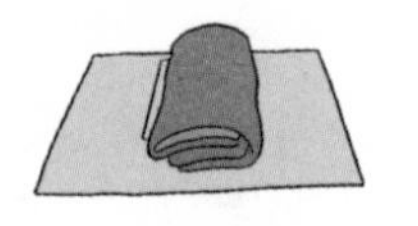

Place the loosely rolled pants on a flat sheet of tissue paper.

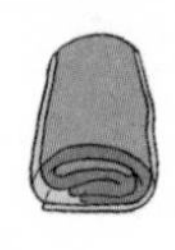

Carefully wrap the rolled garment in the tissue paper.

SWEATER OR LONG-SLEEVED TOP

Lay the garment on a flat work surface, face down.

Lay a piece of tissue paper on top of the garment.

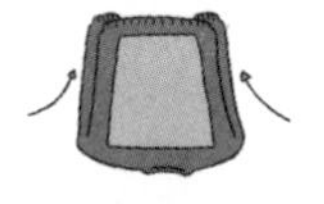

Pull down both sleeves to the sides of the garment.

Fold the sleeves and sides of the garment inwards so they lie on top of the tissue paper.

Bring the hem of the garment ¼ to ⅓ of the way up towards the top of the garment. Fold again until you reach the collar.

Flip the garment over and smooth out any wrinkles.

DRESS SHIRT OR BLOUSE

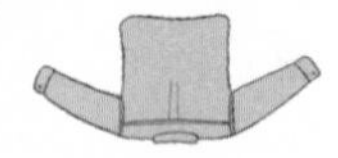

Lay the garment on a flat work surface, face down.

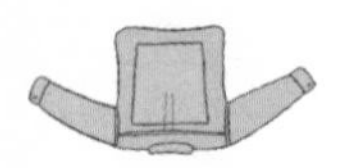

Lay a piece of tissue paper on top of the garment.

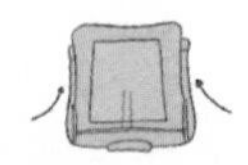

Pull down both sleeves to the sides of the garment.

Fold the sleeves and sides of the garment inwards so they lie on top of the tissue paper.

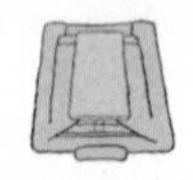

Add another piece of tissue paper on top of the folded sleeves.

Bring the hem of the garment ⅓ of the way up towards the top. Fold again until you reach the collar.

Flip the garment over and smooth out any wrinkles.

Crumple two pieces of tissue paper and insert them into the collar of the shirt. This step is very important because the tissue will hold the collar's shape so that it does not get crushed in the suitcase.

SUIT JACKET (FOLD METHOD)

Lay the garment on a flat work surface.

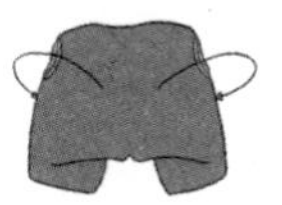

Fold the jacket inside out.

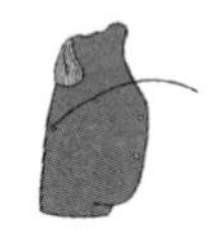

Fold the jacket in half.

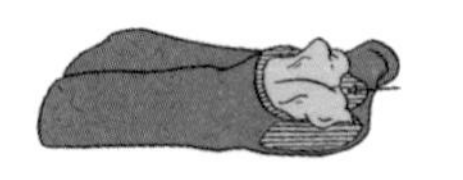

Stuff the sleeves with at least three or four pieces of crumpled tissue per sleeve. This is very important so that the sleeves and body of the suit jacket do not flatten out.

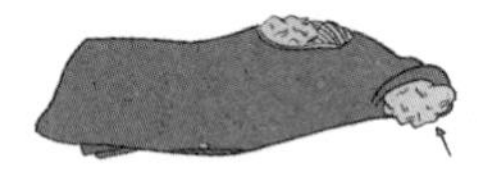

Stuff the collar area with crumpled tissue. This is important so that the collar area does not flatten out.

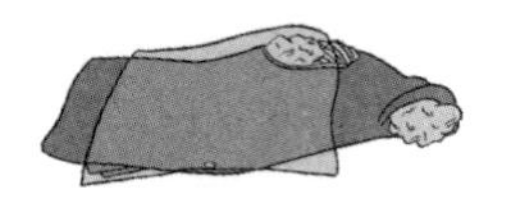

Once the garment is stuffed with tissue, lay another tissue on top.

Gently fold the jacket in half.

MEN'S TIES (FOLD METHOD)

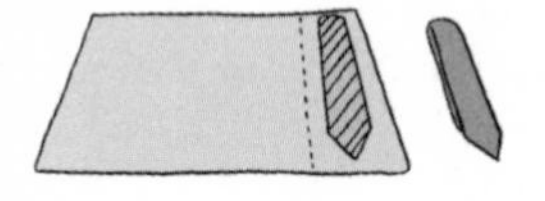

Fold the tie flat. Lay a piece of tissue paper on a flat work surface. Place the first tie on top of the tissue paper.

Fold over the tie and tissue paper so that the tie is now upside down and sandwiched between layers of tissue paper.

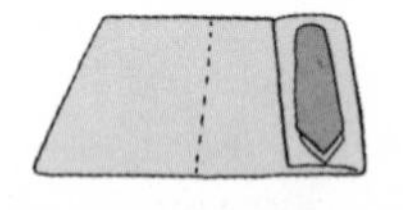

Place the second tie on top of the tissue paper so that the ties now lie back to back.

Grab both ties and tissue paper, and fold over again so that both ties are now sandwiched between the tissue paper.

Repeat again with a third tie, or continue to fold over until you reach the end of the sheet.

DRESS

Lay the garment face down on a flat work surface.

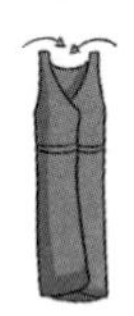

Fold each side of the dress in ¼ of the way.

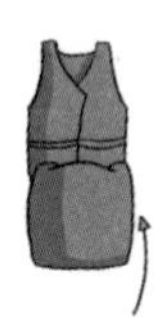

Fold up the bottom half of the dress until you reach the top. The dress is now folded in half.

Flip over and lay flat in the suitcase.

Casual day dresses that are wrinkle-free can be easily folded and packed in your suitcase. For more formal dresses, I would first place the garment in a plastic dry cleaning bag and then use tissue paper between the folded layers.

GOLDEN RULES OF HOW TO PACK A SUITCASE

I cannot overstate how much I like an organized suitcase. Especially on a business trip, why waste time rummaging in a poorly packed bag when, with a little extra care before you leave, you can easily access the precise garments you need, wrinkle-free and ready to wear?

⋄ ◆ ⋄

Before beginning to pack, be sure the lids of any liquids are securely fastened, and place containers inside a sealed plastic bag to prevent leakage. For extra security, consider taping container lids closed.

⋄ ◆ ⋄

Then follow these steps to ensure your clothes arrive in excellent condition.

⋄ ◆ ⋄

Line your suitcase with tissue paper. Place trousers along the bottom of the suitcase with the legs hanging over the outside edge.

Place heavy items such as shoes (wrapped in tissue or other packaging to avoid dirt transfer) along the bottom edge of the suitcase.

Place any rolled or folded items in next.

Place larger folded garments on top. To reduce wrinkles, use lots of tissue paper between layers of clothing, and to stuff up jacket sleeves and around shirt collars.

Wrap the trouser legs over the pile of clothes.

Stuff socks, belts and other compact items into any empty spots.

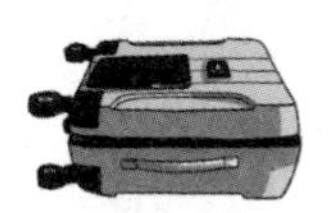

Place your most delicate items in the suit-case last. Review your packing checklist to make sure you've packed everything. Close the suitcase.

For carry-on luggage, have your 3-1-1 bag of liquids easily acessible for when you're going through security.

ENJOYING THE JOURNEY

“I never check luggage. I refuse to check luggage. I look down at people who check luggage.”

DAN LEVY,

Actor, Writer, Producer, TV Personality

Modes of Travel

When you're planning a trip, how you want to travel is just as important as where you want to go. If time is of the essence, nothing beats air travel. However, for those with the luxury of time, travel by sea, rail or road can offer a more leisurely and very enjoyable experience.

AIR

There was a time when flying was a luxury for the elite. Spacious seating, full meal service and a doting service crew—how I yearn for the glamour days of air travel! Things have changed, but there is a lot to be said for the convenience of modern air travel: you can now fly to the other side of the world in roughly fifteen hours on a non-stop flight.

BOOKING

With scores of online booking websites and apps, booking your own flight has never been easier. I may be old-fashioned, as I still prefer to employ the expertise of a travel agent who can navigate the rules and regulations for me (see page 10), but if you have the time and inclination, booking a flight yourself can be a quick and economical option.

BUTLER'S TIP: FREQUENT-FLYER PROGRAMS

Rewards programs are not always straightforward. When you first join, you may dream of all the wonderful faraway destinations you will visit when you redeem your loyally accrued points. Sadly, it's not usually so simple. Here are some tips to maximize your benefits:

Read the fine print. Every program is different, so understand the terms of the program before you sign up. Check to see if there is a membership fee, and be sure to understand the conditions for redeeming points before you start planning.

Be loyal. Pick your airline carrier and stick with it. If your airline is partnered with other carriers, so much the better, as partnered programs will maximize your earning power.

Book early. Book your ticket as soon as you've earned the required number of points. The further out you book, the better your chances of getting what you want.

Flexibility is key. Be prepared to take a more roundabout route. Trying to redeem points for a direct ticket from Toronto to London is almost impossible, but if you are prepared to fly via Ottawa or Halifax your chances of booking that free ticket increase exponentially.

SEAT SELECTION

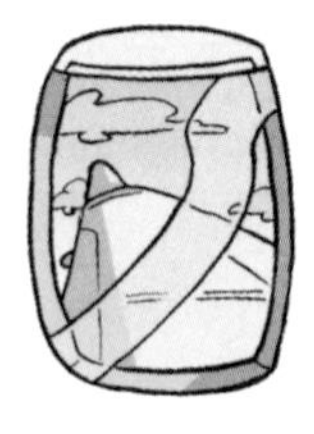

Reserve your seat as far in advance as you can. For peace of mind, I'll pay the fee for seat selection to ensure that I don't get stuck with the dreaded middle seat. Here are a few considerations:

- The least motion during a flight is in the centre of the aircraft, and the most is in the back.
- Bulkhead seats may have more legroom, but they also have much less bin space and no under-seat storage; this can be annoying. They are also often much narrower than regular seats because the tray table and entertainment unit are stored in the arm-rest instead of in front of the seat.
- Seats in an emergency exit row and the row in front of it often do not recline, for safety reasons. Generally, seats in the last row of the aircraft do not recline either.

CHECK-IN

I believe the best way to engage with airline check-in agents is to be polite and friendly (which goes for all customer service personnel you interact with on your travels). Always have your documents ready before you approach the check-in desk, so that the process is as efficient as possible. This will put you in everyone's good books.

BUTLER'S TIP: UPGRADES

Today, the ultra-elite members of any loyalty program have the best chance of getting the elusive upgrade. For the optimist, here are some tips that could improve your chances of being considered:

Dress neatly: anyone dressed in pyjamas or beach attire will not stand a chance.

Check in early and politely ask about upgrade opportunities.

Travel alone, as upgrades are usually given to solo travellers.

Alternatively, several airlines offer meal upgrades for an additional fee. If you're seated in economy but are yearning for a more business-class dining experience, check your airline's website. These meals must be ordered and paid for in advance.

DID YOU KNOW?

In the 1960s, Pan Am became the first airline to introduce freshly baked cookies on board their flights.

Make sure your checked luggage is packed within the weight limits before you get to the agent at the counter. Airlines are strict today about luggage weight. If you're an over-packer, weigh your luggage at home, or make a point of finding a luggage scale at the airport prior to check-in. If your luggage is overweight, do not inconvenience the agent and the other people behind you by unpacking your bags at the counter. Try to cram those extra-heavy items into your carry-on before approaching, and if that is not an option, courteously accept that you'll have to pay the excess baggage fee.

AIRPORT SECURITY

How often have you stood in a security clearance line that feels as though it will never end, stuck behind someone who is totally unprepared and holding up the line? Here is how to move through security faster:

- Have your boarding pass and passport ready for verification.
- Dress so that your coat or jacket, belt and shoes are easy to remove.
- Avoid clothing and accessories with metal, as they may set off the security alarm.
- Remove your large electronic items, such as your laptop computer and tablet, from their cases and place them in the bins provided.
- Pack your liquids in your carry-on according to the 3-1-1 rule (page 57) and make sure they're easy to access when you go through security.
- Gather all your belongings quickly once they've passed through the scanner, and then move to one side to repack items, replace your shoes, etc.
- Be polite to the security personnel.

DUTY-FREE

I have never been much of a duty-free shopper myself, but it seems airports today have become the new shopping malls. If you decide to engage in some duty-free shopping:

- Give yourself plenty of time to shop. There is nothing more rude than shoppers who are impatient for service because their flight is about to depart.
- Have your passport and boarding pass with you. You will need to present them at the time of purchase.
- Don't be alarmed by the aggressive sales clerks. Bear in mind that they often work on commission.
- Always check ahead of time what you can and cannot bring home. Each country has different customs allowances for goods, tobacco and alcohol. If you exceed your allowance, be prepared to pay customs duty.
- Keep your duty-free purchases with itemized receipts in their official security bags; keep these bags sealed.
- Flights are not BYOB. Aviation regulations and federal laws prohibit the consumption of your own alcohol on board the plane.

DID YOU KNOW?

An Irishman by the name of Brendan O'Regan invented the concept of the duty-free shop in 1947.

FLIGHT DELAYS AND CANCELLATIONS

Unexpected delays or cancellations, along with lost or damaged luggage, are the bane of any traveller's existence. While it's all too tempting to vent your frustrations, don't! Remember, your fellow passengers are all in the same boat, so remain calm (and count to ten, if that helps).

If you need to rebook your flight, try calling the airline while you also wait in line to talk to the agent in person. And be flexible with the available options. If it's late and I'm definitely stuck, I'll choose to

BUTLER'S TIP: BE COURTEOUS

Travel can be frustrating, but that is no excuse for rudeness. We should always remember to practise good manners, wherever and whatever the situation. Treat security agents, airline check-in agents and all customer service personnel with respect and politeness. Being courteous goes a long way.

spend the night at a hotel. At least I'll get a good night's sleep and feel ready to tackle the next day.

MISSED CONNECTIONS

There is almost nothing worse than missing a flight connection, and the stress and disruption it can add to your journey. Flights often get delayed, so if you have a tight connection, politely mention it to the flight attendant when you board. With luck, they will help you get off the plane first. Failing that, politely explain your circumstances to the passengers in front of you and ask for their assistance, and they may let you pass ahead of them.

BUTLER'S TIP: HOW TO AVOID GETTING BUMPED

Overbooked flights seem to be the norm these days, as airlines do their best to minimize the impact of last-minute cancellations. So, how can you reduce your chances of getting bumped?

Passengers with the cheapest tickets will get bumped first, so if you thought you got a deal on that ticket, think again.

Passengers who arrive late will usually get bumped. Check in online in advance, and arrive early at the airport.

Reserve your seat when you book your flight so that it is guaranteed. Depending on the cost of the ticket, you may need to pay an extra fee to pre-select your seat.

Family groups and minors travelling alone generally will not get bumped.

HOW TO FLY LIKE A BUTLER

Eat something before you board. Never start your flight hungry. Avoid eating junk food before a long flight. Greasy and salty foods are even more difficult to digest when you're in the air.

Always bring snacks with you for convenience, and in the event of flight delays.

If you would like or need a special meal, make sure you pre-order it. And be prepared to eat what you are given—the general meal service is no longer an option for you, even if it looks better. There is no guarantee that you will get your special meal so be flexible if there is a mix-up.

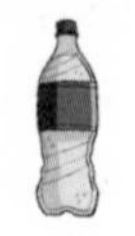

Always stay hydrated. This is one of the most important things to remember!

Be careful with caffeine. It is a diuretic and will dehydrate you. Herbal tea is a great substitute, or hot water with lemon slices—most airlines have lemon on the drinks cart.

Avoid drinking too much alcohol. It's also a diuretic, and some experts say it prolongs jet lag.

Fewer airlines offer pillows and blankets so consider bringing your own, especially on long-haul flights. Remember to keep within your seat space; don't lean your pillow against your neighbour. When you are done, neatly put them back in your bag.

Do discreet in-seat exercises such as ankle circles, knee lifts or neck and shoulder rolls every hour to ensure good blood circulation.

Don't forget all your power cords and travel adapters so that you can stay connected. Bring your own headphones as well, as airlines rarely provide them free of charge.

Always wear a pair of comfortable closed-toe shoes. These are especially important in an emergency when you need to get off the plane.

DID YOU KNOW?

In 1987, American Airlines saved $40,000 by removing one olive from each salad served in first class.

AIRPLANE ETIQUETTE

Does the overhead bin belong to the person seated below it? Technically no, but be respectful. Everyone is entitled to some space in the overhead bin, and it is polite to use the bin located closest to your own seat.

If I am sitting in the middle seat, who gets the armrest on either side of my seat? As the middle passenger, you get the inside armrests, as the passenger on either side has an outer armrest.

What kind of food can I bring on board to eat during my flight? Since meals are rarely served on flights now, and the buy-on-board options are generally not very good, it is common for passengers to bring their own food on the plane. Consider your fellow passengers, and select a meal that creates minimal smell and mess. Sandwiches or wraps make ideal choices, since they can be eaten unheated and without cutlery.

How do I deal with the person sitting in front of me who fully reclines his seat? This can be frustrating but all passengers do have the right to recline their seat. During meal service, it is acceptable to ask the passenger in front of you to put his seat upright. And if you are uncomfortable asking, politely mention this to the flight attendant, who will ask on your behalf.

The child behind me keeps kicking the back of my seat. What should I do? The most important thing is to stay calm and be as empathetic as possible. Address the parents politely and say something like, "Do you mind asking your child to stop kicking my seat, please? Thank you. I'm trying to have a nap."

How do I handle the chatty passenger seated beside me?
I remember getting great advice on this very topic from the late Letitia Baldrige, who advised saying something like, "It is wonderful to meet you, but if you don't mind, I really want to get back to my book. It's so riveting, and I want to keep reading!"

A family has asked me to swap seats with them so that they can all sit together. Do I have to give up my seat? The short answer is no, but put yourself in the other person's shoes. It's always nice to sit with someone you know.

My family are seated separately but we would rather sit together. How do we best arrange to change seats? Approach the gate agents to see if they can help. If they can't, then ask a flight attendant upon boarding. If you need to ask a fellow passenger to move, remember that you are asking for a favour, and that they may have paid extra to secure that seat. Politely and sincerely ask, "Do you mind?" or "Could you help us, please?" You will find most travellers will be accommodating.

How do I navigate around a sleeping passenger? As tempting as it is to try to climb over someone without disturbing them, it is always best to tap them gently on the shoulder and excuse yourself. You wouldn't want the passenger to wake up as you're straddled over them, struggling to get out to the aisle.

Is it appropriate to tip a flight attendant if they go above and beyond the call of duty? No; as Amy Vanderbilt put it: "No member of a plane's personnel is ever tipped. One says good-bye to the crew in attendance at the jet bridge when debarking."

THE ETIQUETTE OF AIR TRAVEL

If you're going to remove your shoes, always wear socks, and be sure to put your shoes back on when walking throughout the cabin.

⬩◆⬩

Stay within your limited seat space as much as possible.

⬩◆⬩

Silence is golden. Other passengers may be trying to work or nap on the flight.

⬩◆⬩

Keep hydrated with water and remain sober.

⬩◆⬩

Clean up after yourself at your seat, and in the lavatory.

⬩◆⬩

Watch over your children and keep them in check.

PRIVATE AIRPLANE

If you are fortunate enough to be invited on board a private jet, here are some things you should know:

Guest "Do"s	Guest "Don't"s
Dress appropriately. Unless you are travelling with immediate family, business or business casual is your best bet.	Do not forget your passport or any other required travel documents.
Wait for the host before you take a seat.	Do not be late, and never keep your host waiting.
Obey safety rules; wear your seatbelt.	Do not ask the host or crew inappropriate questions, such as the cost of the flight.
Be on your best behaviour. Sit quietly during the flight, and drink alcohol only in moderation.	Do not expect the same service you would receive on a commercial aircraft.
Keep yourself busy, and bring your own reading material or personal electronic device.	Do not travel if you are sick. You do not want to risk infecting your host and other passengers.
Limit your luggage.	Do not wear strong scents on board the flight as they can be overwhelming.
Declare everything you purchase. Flying privately does not make you exempt from customs regulations.	Do not brag about your experience by taking photos and posting on social media.
Send a thank-you note to your host after the trip.	Do not bring a gift for the host; there is limited space on board.

DID YOU KNOW?

Singapore Airlines' Business Class seats remain the widest in the industry, at 28 inches. Singapore Airlines is also the first to operate "Green Package" flights featuring the use of sustainable biofuels, fuel-efficient aircraft and optimized flight operations.

RAIL

Train travel has never been as popular an option in North America as it is in Europe and Asia, and I suspect that has something to do with the long distances between cities. Nonetheless, the romance of train travel is very appealing to me. The journey may take longer, but watching the landscape change through the window as you roll through the countryside is mesmerizing!

Whether you are travelling on a transcontinental or any other type of train, here are some guidelines to follow to help ensure that everyone has a good trip.

DID YOU KNOW?

The Trans-Siberian Railway is the world's longest single railway line, spanning 5,772 miles and crossing seven time zones.

Should I offer my seat to pregnant women, older people and those with disabilities? I rarely see this happen any more. Chivalry is not dead! If you are sitting in a carriage of unreserved seating, then the younger and more able-bodied should always offer their seat to others who need it more than they do. It's the right thing to do.

Am I required to tip during train travel? Whom do I tip, and how much? It's not required, but service crew personnel might be tipped based on the quality of service received. Consider $10-$20 for the car attendant, $5-$20 for the sleeping car attendant, 10 per cent for the snack bar attendant and 15 per cent for dining car staff. Train station porters should be tipped $1 per bag.

Is the Orient Express still as formal as it once was? The Venice Simplon-Orient-Express is still the epitome of luxury train travel. The journey on board vintage carriages has retained the romance, glamour and service of a bygone era. Smart and formal attire is de rigueur.

What can I do when someone is speaking loudly and is rudely bothering everyone in the train car? A crowded train car, in my opinion, is most likely not a good place to confront someone regarding their behaviour. Try to stay focused on your book or electronic device, and hopefully the disruption will come to an end quickly. If you do need to bring it up with someone, always speak discreetly to a train official. They are trained to deal with these situations.

THE ETIQUETTE OF TRAIN TRAVEL

Silence is golden. Always use your "inside voice," and keep the volume down on your electronic devices.

◆ ◆ ◆

Unless appropriate, do not sit in seating reserved for the elderly, those with disabilities and pregnant women.

◆ ◆ ◆

Whatever food you bring on board the train should not smell, and should be easy to eat quietly.

◆ ◆ ◆

Seats are for people; keep your feet, shoes and bags on the floor.

◆ ◆ ◆

Keep the aisles clear.

◆ ◆ ◆

Let passengers with bags, children and food pass first when you are in the aisle.

◆ ◆ ◆

Keep track of your luggage, and never let your bags out of your sight.

CAR

North Americans love to travel by car. I personally love a good road trip, especially when we all agree on who is going to drive and the type of music to listen to!

So, how do you plan a successful road trip? Well, first you need to decide if it is going to be a leisurely journey with stops along the way, or a mission to get as quickly as possible from point A to point B. This will dictate the route you will take and the pace of your drive.

DID YOU KNOW?

In Japan, it is extremely rude to talk on your cellphone while on the train. Also, passengers are reminded to set their phones to mute, which in Japan is called "Manner Mode."

When I used to drive frequently between Toronto and New York City I found that the eight-hour drive went by relatively quickly if I was on the road no later than 6:00 a.m. The early start time meant that there was less traffic to contend with, and by the time I arrived at 2:00 p.m. I had the rest of my day to enjoy the city.

If we take my car on a road trip, is it fair for me to ask others to share the driving? Perfectly fair. The key is to discuss expectations ahead of time, so that everyone is in agreement before you set out.

When I drive my friend's car on our road trip, she is always complaining about how fast I am going. I tell her when I drive it's my way, and when she drives it's her way. Is this fair? In this case, you must respect your friend's wishes: her car, her rules.

My children often fight on road trips. How do I deal with this? Plan ahead and have activities on hand to keep them entertained. Videos, video games and tablets are options, but don't forget crayons, paper, colouring books, car games, toys and books. And plenty of snacks.

Should I drive my own car or rent a car for a road trip? Well, it depends. Does your car meet your family's needs, and is it in good condition? If the answer is yes, then save the cost of a rental car by using your own.

THE ETIQUETTE OF ROAD TRIPS

Decide who will be the designated driver(s) and set a driving schedule ahead of time.

◆ ◆ ◆

Music choices can be contentious, so agree on a protocol in advance. Will one person play DJ for everyone? Will you take turns selecting the music? Is talk-radio an option, or audiobooks? Or will you maintain silence in the car so that passengers can listen to whatever they like with their headphones on?

◆ ◆ ◆

Remember that a car is a very small space. Passengers may become uncomfortable or irritated during long drives, so try to be patient with others. Bring something to occupy yourself, and don't rely on others to entertain you.

◆ ◆ ◆

Pack your car with essentials such as food, snacks, water, napkins and wet wipes.

◆ ◆ ◆

Pack a pillow and blanket for those who want to sleep.

COACH

Whenever I think of travelling by coach, I fondly recall my high school trip to Italy. Touring the country by coach was a fantastic experience, and we were even lucky enough to attend the Wednesday Papal Audience with Pope John Paul II. But coach travel also makes me think of the movie *If It's Tuesday, This Must Be Belgium*, which captures the whirlwind experience of travelling by coach perfectly. It can be fun, but it can also be exhausting, and no matter how much you like the people you are travelling with, they will probably get on your nerves.

If you are on a budget, travelling by coach is a great option, and coach tours offer many benefits for the first-time traveller to an unknown destination. Coach travel also provides a safe way to meet new people and make friends from other parts of the world. And it's also the easiest way to see the sights, especially if you don't speak the local language. Experienced guides will tour the group and offer advice on the best way to experience your destination. And since tour operators usually have pre-arranged priority entry, you can skip the lines.

Is a coach tour a good idea for a family with children? As with car travel, your children will be confined to their seats, but on a coach you do not have the flexibility of taking breaks when it suits them. Look into child- or family-friendly tours that will cater specifically to their interests. That way, you'll be travelling with like-minded passengers.

Does a coach tour bus offer any amenities? Coach tour buses are air-conditioned and often have reclining seats with generous legroom. There is usually a washroom on board, and most have free Wi-Fi.

CRUISE

Transcontinental leisure travel was once dominated by large ocean liners. When air travel became more popular and affordable, ocean liner travel declined, and in came the cruise ship. Cruise ships travel to several destinations in one voyage and offer an array of onboard experiences to appeal to the interests and personalities of as many travellers as possible. Today's full-service cruise ships (remember *The Love Boat*?) are huge vessels, offering more amenities and entertainments than you can imagine.

As appealing as this sounds, there are a few disadvantages to cruising: cabins are notoriously small, you will be sharing a table for meals, and there is always the possibility of seasickness and bad weather.

How should I dress aboard a cruise? Follow the dress-code guidelines of the ship. As I heard it said once, "Dress for a cruise, not a car wash!" A bit harsh, but the point is to dress appropriately—you certainly don't want to make yourself or others feel uncomfortable.

Do I need to bring a jacket and tie? It is always good practice to have one in your suitcase. You may not use it, but if you end up being invited to the captain's table, you will be prepared.

What are my responsibilities towards the other guests at my dinner table? First, introduce yourself to your new dining companions. Be polite, smile, acknowledge everyone, listen, participate in conversation and mind your table manners.

What if I don't like the guests at my table? No matter what, you must be polite. However, if you do not think you can continue to be seated at the same table, speak privately with the maître d'hôtel and have him change your table assignment, preferably to a table far away from your former dining mates to avoid awkward encounters in future.

How much, whom and when should I tip while on board?
Tipping standards vary depending on the cruise. When you book your cruise, check the terms and policies. Some cruise ships include the tip in the fare or allow you to charge your tips to your onboard account. If an employee goes above and beyond the call of duty, an extra gratuity is always appropriate. In this case, give the tip discreetly to the employee in an envelope (use the ship's tip envelopes if available). The crew members to tip are the butlers, cabin stewards and dining room staff.

DID YOU KNOW?

Cruise ships are equipped and staff are trained to fend off pirate attacks.

THE ETIQUETTE OF CRUISE SHIPS

Pay attention to the safety demonstration.

⁘

Be aware of others; as large as a cruise ship may be, it is still a confined space.

⁘

Follow the dress codes.

⁘

If you are feeling ill, return to your cabin.

⁘

Don't be late returning from your visit on shore; make sure you're back on board at the required time.

⁘

Deck chairs are for everyone. Don't be a hog.

YACHT

If you are planning to travel by private yacht, lucky you! It will be an incredibly luxurious experience. The service is generally exceptional. Though the cost would set many of us reeling, if you have the means it's well worth it.

How do I charter a yacht? It is best to find an established charter broker who will match a yacht to your needs and expectations. If you're not sure whom to contact, refer to professional organizations such as MYBA The Worldwide Yachting Association, which has a list of vetted members. The broker will do an in-depth review of your requirements and present several options to you based on the information you provide.

How do I check references? If possible, talk to previous clients. Have a list of questions prepared in advance and make sure they are open-ended so that you get meaningful and detailed answers. For example, instead of asking "Do they handle dietary restrictions?" you might ask "How did they handle special dietary restrictions?"

What do I do if I'm not happy with the service? Always speak to the captain if there is a problem, and address issues as they arise so that they can be dealt with immediately. Remember, be reasonable and courteous in your communication (page 87).

THE ETIQUETTE OF YACHT TRAVEL

Pack your belongings in soft-sided luggage, which takes up less space in storage.

◆ ◆ ◆

Respect the "bare feet rule." No shoes (unless they are deck shoes) are allowed on deck or in the cabins. Place your shoes in the basket conveniently placed at the entrance to the deck.

◆ ◆ ◆

Be mindful and respectful of boating safety measures, such as life jacket and lifeboat procedures.

◆ ◆ ◆

Treat the crew respectfully and allow them to do their job. While their job does include guest services, it does not include babysitting—you, or your children.

◆ ◆ ◆

The yacht's back-of-house and the crew's private areas are off-limits to guests. Unless you've been invited, do not enter their space.

◆ ◆ ◆

Speak to the captain if you have
any requests or concerns.

◆ ◆ ◆

Be on good behaviour. There is
zero tolerance for illegal activities.

◆ ◆ ◆

Give the chef advance notice if you're planning
to skip a meal or dine ashore, so that
food and time won't be wasted.

◆ ◆ ◆

Tipping is customary. At the end of the trip,
give your tip to the captain, who will distribute it among
the crew. The industry standard is 10 to 20 per cent
of the charter fee, but feel free to give more if
the experience was exceptional.

On Arrival

YOU'RE ALMOST THERE! You've arrived at your destination, and now all you need to do is clear customs and immigration, collect your bags and be on your way!

IMMIGRATION

It is the job of immigration officials to check that you are who you say you are, and to determine if you should be allowed entry into their country. If they are intimidating, that's likely intentional, so take a deep breath and don't take their tone personally. Answer questions simply and honestly, and allow them to do their job.

CUSTOMS

My best and only advice when it comes to clearing customs is: declare, declare, declare. There is absolutely no advantage to hiding and lying about your purchases just to save a few dollars. Getting caught and having to pay the penalty is far worse. A record will be kept of your infractions, meaning that you will be flagged and may undergo more detailed inspections in the future. Be honest, and you'll find the customs agents are generally very reasonable.

MISSING LUGGAGE

It's every traveller's worst nightmare: landing at your destination only to find your luggage is missing. I'm happy to say that with all of the miles I have flown I have lost my luggage only twice, and in both cases I was fortunate to have it recovered quickly. I think that I have been lucky with my luggage in part because I observe a couple of hard-and-fast rules. I always go to the airport early to check in—usually two hours ahead, and sometimes more. And I always leave plenty of time between connecting flights, because it gives baggage handlers enough time to transfer bags.

But what should you do if it does happen? First, report your loss to the airline's baggage service counter before you leave the airport. Make sure you have your baggage-claim receipt, and be prepared to give a detailed description of your luggage. If you have a photo, even better. In return you'll be given a file number or telephone number to track the status of the search. Usually lost luggage is located and returned within 24 hours.

BUTLER'S TIP: PHOTOGRAPH YOUR LUGGAGE

I always take a photo of my luggage with my cellphone before checking it, so that if it gets lost I can show the baggage service attendant exactly what it looks like.

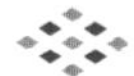

MEETING AT THE AIRPORT

There is great joy and excitement in meeting loved ones at the airport. For a problem-free reunion, be sure to prepare ahead of time. Get the travellers' flight details, and be sure to share cellphone numbers in case you have trouble finding each other. Set up a Plan B so that if you are unexpectedly unable to meet they will know what to do. On the day of the flight, check with the airport or airline to ensure the flight is on time. When you meet, have a luggage trolley ready. And be prepared for your travel-weary loved ones to want to unpack and get settled in before you plan a night on the town.

TRANSPORT TO YOUR ACCOMMODATION

When you are planning your trip, it's important to determine ahead of time how you are going to get to your accommodation. Whether you choose a car service, a taxi, car rental or public transit, do your research, and find out the price in advance so there won't be any surprises.

BUTLER'S TIP: CAR RENTAL

Think about what your needs are and what your budget is. This will help you decide what kind and what size of car you want to rent.

⬩◆⬩

Always book ahead of time; never leave it to the last minute in the hopes of a better deal, because selection will usually be much more limited.

⬩◆⬩

Car insurance is very important! You may have a credit card that will give you coverage, but check the fine print to make sure it will be enough. Personally, this is one area I won't scrimp on, because you never want to be short on collision or liability insurance.

Method of Transportation	Pros	Cons	Summary
Car Service	Offers safe, comfortable transportation to wherever you are going.	Quite costly.	Best suited to trips where you have important and time-sensitive engagements to attend.
Taxi	Requires the least planning in advance.	Can be fairly costly, especially if your destination is not close to the airport.	Taxis are usually the best option if your destination is close to the airport, or you are unsure of your arrival time or travelling at an unusual time of day.
Car Rental	Offers the most flexibility; you can drive to any destination.	It can be overwhelming to drive in an unfamiliar city. And you'll need to worry about keeping the car in pristine condition.	A rental car is a good choice if you plan to do lots of driving and will need the car throughout your trip.

Method of Transportation	Pros	Cons	Summary
Local Transit	The most cost-effective option.	Public transit can be an inefficient and confusing way to reach your destination, especially if you don't speak the local language.	Public transit is useful in large cities with well-developed metro systems, and/or for budget travellers who aren't on a tight schedule.
Hotel Airport Shuttle	There's lots of room for luggage, and often this service is included in the cost of the hotel.	The shuttle is available only for select hotels, and often runs on a fixed schedule.	A hotel shuttle is the best option if you are travelling directly to a hotel that offers this service.

DID YOU KNOW?

The Shanghai Maglev train is the world's fastest train, travelling at a speed of 267 mph from Shanghai's Pudong International Airport to Longyang Road Station in suburban Shanghai. The 19-mile distance takes only 7 minutes to complete.

DURING YOUR STAY

"There are two kinds of cruises—
pleasure, and with children."

GEORGE BURNS,

Comedian

Checking In

ARRIVING AT YOUR DREAM destination is the moment when your trip truly begins. Whatever anxiety and stress you might have been feeling vanishes as excitement sets in.

ENJOYING YOUR HOTEL

I prefer to stay in hotels when possible, and what follows are my tips for making the most of your hotel and its amenities. I like to spend a few minutes first unpacking my suitcase, setting up my toiletries in the bathroom and figuring out how the TV and thermostat work. Once I've familiarized myself with my new surroundings, I can truly relax.

BUTLER'S TIP: STAYING IN HOTELS

Bring your own toiletries. They will nearly always be better than those provided by the hotel.

◆ ◆ ◆

Freshen the air in your room. If it is the dead of winter and your hotel room does not have a window that opens, the air is often very dry. I fill my bathtub halfway with hot water and leave the bathroom door open. This really helps to humidify the room.

◆ ◆ ◆

Bring your own entertainment. A good book you've been meaning to read and a glass of your favourite beverage provide just the right touch when you need to relax and recharge before another busy day.

◆ ◆ ◆

Take a few minutes to read the hotel directory in your room. It is filled with great information and can help you make the most of your stay.

◆ ◆ ◆

Don't be afraid to ask. If you don't like your pillow, the room is too cold, or it is just too noisy near the elevator, ask for assistance.

SECURING THE BEST ROOM

Not all hotel rooms are created equal. When your objective is to get the best hotel room available, or possibly an upgrade, keep these tips in mind. And whatever happens, smile and be nice!

- Join the hotel's loyalty program to collect points towards upgrades and other benefits. Elite-status members get preferential treatment.
- Call the hotel directly to book your reservation. Often reservations booked through an online discount site get lower priority and are less likely to be assigned the best rooms.
- When booking your reservation, request a room that is away from the elevators, ice machines and other noisy, high-traffic areas.
- Check in early while room selection is good.
- Inspect the room before you unpack. If you need to change rooms, make the request as soon as possible.

CHANGING ROOMS

If you do need to change rooms, be friendly, respectful and courteous when you are making the request. Have a reasonable explanation, and be specific with details: noise, odd smells or unsuitable bed configurations are

reasonable reasons to request a change. Speak with the front desk staff, who will handle your request. Be calm and patient, and manage your expectations: the hotel will be working under the constraints of availability.

CONCIERGE

The concierge is the hotel's oracle, knower of all things, and an invaluable resource. The job of the concierge is to make sure hotel guests are well looked after. Among other things, they will happily make reservations, secure tickets, book transportation or offer sightseeing and dining suggestions. Don't be shy to make unusual requests, as long as they are reasonable and legal. If you can, give the concierge some notice when making your request to increase the likelihood of success.

BUTLER'S TIP:

Remember to thank the concierge and tip to show your appreciation. Tip well if they have gone above and beyond. I prefer to tip with each request rather than at the end of my stay.

UNPACKING

To unpack or not to unpack? It's personal preference, and will depend on the length of your stay. I generally unpack and hang wrinkle-prone garments, but keep my packing cubes as is. Here is my routine:

1. Put passport, money and other valuables in the hotel safe.
2. Open your suitcase in an out-of-the-way spot.
3. Hang all items that need to be as soon as possible.
4. Put the toiletries bag in the bathroom. Pull out toiletries and have them rest on a face cloth placed on the vanity.
5. Leave everything else in the luggage.

BUTLER'S TIP:

Leaving clothing and personal items scattered all over the room is a sure way of leaving something behind. Always put back what you pulled out from your suitcase, or place items in one specific spot in the room.

LAUNDRY ON THE ROAD

Doing laundry on the road is a good way to minimize what you need to pack. However, most hotels prefer that you do not do laundry in your room. I will hand wash the small, simple essentials, but I am always careful and respectful of the room.

HAND WASHING LAUNDRY

1. Rinse out the bathroom sink to remove any residue. Fill the sink with water (cool or warm depending on the garment) and add a drop or two of laundry soap (if you have it) or shampoo.

2. Swish the items through the soapy water until clean. Drain the sink and squeeze the items to remove the excess soapy water. Refill the sink with clean water.

3. Rinse garments thoroughly. Rinse again if there is still soapy residue. Squeeze the items to get rid of the water.
4. Lay the items flat on a clean towel and then roll the towel up from one end. Squeeze the roll to remove as much excess water as possible.

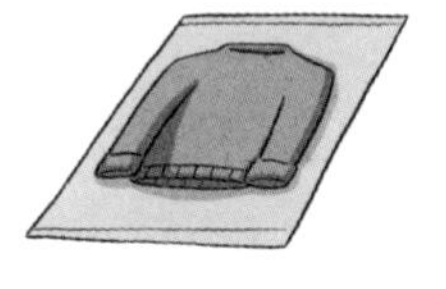

5. Unroll the items from the towel. The items should no longer be dripping wet at this point. Hang or lay the garment on a fresh towel to dry.

STEAMING

Most good hotels can offer this service for you, for a price. Or you can try steaming your clothing yourself in the bathroom. Run a hot shower, then hang your wrinkled suit or dress on the rod but away from the shower stream. Be aware that condensation from over-steaming might ruin the garment, so don't let the shower water run too hot or too long. Keep an eye on the garment until most of the wrinkles are gone.

GOLDEN RULES FOR AVOIDING JET LAG AND ADJUSTING TO TIME ZONES

No matter how seasoned a traveller you are, jet lag is a fact of life. There is no simple solution—the key is to adapt to the local time zone as soon as possible.

◆ ◆ ◆

If you are on the night flight from North America to Europe: **Try to have a meal on the ground before you depart. Then have a drink on the plane after takeoff to relax.**

Get some sleep. Put on your noise-cancelling headphones and your eye mask. It will take effort, but force yourself to sleep, and try to ignore the rest of the plane. This is where the window seat is helpful.

Stay awake on arrival. When you arrive at your destination, shower, change your clothes and force yourself to stay up. Have a maximum two-hour nap in the afternoon, followed by an early dinner and bedtime. This always works for me.

⬩◆⬩

If you are on the day flight from North America to Asia:
Stay awake on the flight. Watch a movie, read, doodle, anything that will keep you from drifting off.

Sleep on arrival. When you get to your destination, have a light dinner and go to bed.

Melatonin—a natural hormone supplement that encourages sleep—works well for some people. If it works for you, you are one of the lucky ones! Sadly, it does nothing to help me.

Making the Most of Your Time

WE NEVER HAVE AS MUCH time to explore a new destination—or to enjoy revisiting an old favourite—as we would really like. That's why it makes sense to think ahead, make a plan, and also familiarize yourself with local customs and etiquette, so that your trip goes as smoothly as possible. You want to leave with fond memories, and to leave fond memories behind with the people you meet, as well!

ITINERARIES

When I was a butler, getting the family ready for their travels included the preparation of their itineraries.

There were two types. One was a daily itinerary, which outlined in detail all the scheduled events for the day. The other was the travel itinerary, with all their flight and other travel-related details.

There is a certain amount of work and research involved in making these itineraries prior to any trip but it's time well spent. Include names, addresses, phone numbers, confirmation numbers, driving directions and all other pertinent information. That way you'll be able to hit the ground running once you've reached your destination, and have all the pertinent details at your fingertips.

Here are two examples of how a butler prepares an itinerary.

PARIS TRIP

Daily Itinerary for Mr. and Mrs. Jones

Tuesday, 9 October 2018

08:45 **Car pick-up Ritz Hotel** Driver: Pierre +33 607 696 747
Main Entrance
Going to Chanel
31 Rue Cambon Chanel: +33 1 44 50 66 00

09:00 **Arrive Chanel**
Private shopping appointment
Sales Associate: Mme. Hebert Cell: + 33 594 999 023

12:30 **Car pick-up Chanel**
Going to Fouquet's Restaurant
99 Av. des Champs-Elysees

13:00 **Arrive Fouquet's**
Reservations: +33 1 40 69 60 50
Confirmation Number: 12654327
Reservation Name: Andrew Jones—four guests
Meeting: Mr. and Mrs. Lee
Betty Lee Cell: 561-123-4567
Andrew Lee Cell: 561-765-4321

PARIS TRIP, CONT.

Daily Itinerary for Mr. and Mrs. Jones

Tuesday, 9 October 2018

15:00 **Car pick-up Fouquet's** Driver: Pierre +33 607 696 747
Going to Ritz Hotel
15 Place Vendôme

15:30 **Arrive Ritz Hotel**

19:00 **Cocktails: Bar Hemingway in Ritz Hotel**
Meeting: Mr. and Mrs. Smith
Carol Smith Cell: 212-123-4567
Bob Smith Cell: 212-987-6543

20:00 **Car pick-up Ritz Hotel** Driver: Pierre +33 607 696 747
Back entrance
Going to La Tour d'Argent
15 Quai de la Tournelle

20:30 **Arrive La Tour d'Argent** Reservations: +33 1 43 54 23 31
Confirmation Number: 234318
Reservation Name: Andrew Jones—four guests
NOTE: *Driver to wait for you for return trip to hotel post dinner*

PARIS TRIP

Travel Itinerary for Mr. and Mrs. Jones

Wednesday, 10 October 2018

07:30 **Car pick-up Ritz Hotel** Driver: Pierre +33 607 696 747
Main Entrance
Going to Charles de Gaulle Airport

8:30 **Arrive Charles de Gaulle Airport**

08:45 **Check-in: United Airlines Flight UA56, CDG to JFK**
Terminal One
Reservation Code: ASEO3L
Mr. Andrew Jones Seat 3A
Mrs. Mandy Jones Seat 3B
* *Special Meal ordered for both parties*
* *Lounge on Level C, past security, near Gate 56*

10:50 **Flight Departure**

15:45 **Flight Arrival JFK**
United Terminal

16:30 **Pick-up Post 12 (regular spot)** Driver: John 917-123-4567
Going to 740 Park Avenue

HOW TO GET AROUND

When travelling to a new destination I do my best to familiarize myself with the local transportation options. Taking public transit is often the best way to get about, particularly in large cities. I always took taxis in New York City until someone told me that JFK Jr. used to take the subway, and that it was fast and safe. I tried it and it was exactly that: fast, safe and an awful lot cheaper!

SELF-GUIDED EXPLORING VS. GUIDED TOURS

Having organized itineraries for many others, I find that there are two types of travellers. There is the independent traveller, who prefers the freedom, flexibility and control of their own itinerary, and there's the group traveller, who enjoys social interaction with other travellers, plus the added convenience of having an expert guide handling all the logistics. I personally prefer the independence of travel on my own, but I also love the convenience of having a guide. When I was in India on business, I made time to visit the Taj Mahal. I hired a private guide who took me through the monument and its grounds, the Red Fort and the Agra Fort. What made the trip particularly interesting was not only the knowledge and wealth of information offered by my guide but his passion and enthusiasm. It was definitely the trip of a lifetime!

DINING

If fine dining is an important part of your travel experience, booking restaurant reservations is an important pre-trip task. Unless you already have a bucket list of can't-miss restaurants, do your research: consult food sites and blogs, food magazines and travel guides, and solicit personal recommendations from friends and family.

OpenTable, an online reservation service, is a convenient way to book, but for very exclusive restaurants, plan on booking by phone months in advance. Be polite, and don't even think of lying or name-dropping—it never works. If you haven't booked in advance, your hotel concierge might be able to help. To increase your chances of getting a table, be flexible. A dinner reservation that is too early or very late may not be your first choice, but at least you'll be dining in the restaurant of your dreams.

DRESS CODE

Never find yourself turned away at the door because you didn't bother checking the restaurant's dress code. I was once invited to dine at a restaurant that required men to wear a jacket and tie. In my ignorance I showed up with neither, and I was denied entrance. It was embarrassing, to say the least! So, do yourself a favour, and always call ahead to find out the dress code.

EATING AROUND THE WORLD

It's important to learn the dining etiquette of different countries. Here are a few tips so you don't offend anyone at the dining table.

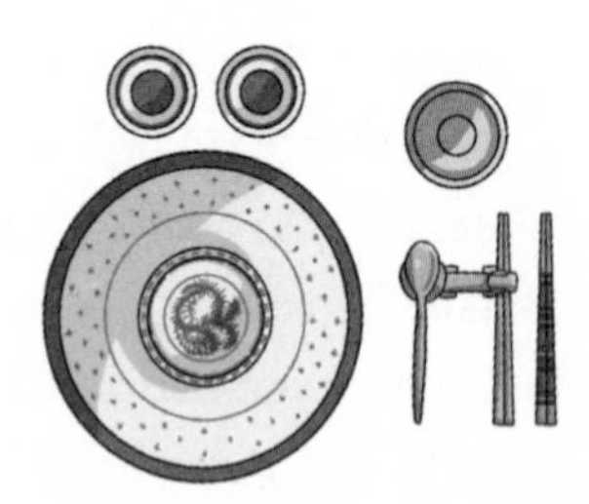

JAPAN

- At the beginning of the meal, use the damp, warm towel given to you to clean your hands only. Save washing your face or neck for bath time.
- Don't ever pass food from one person's chopsticks to another person's. This is offensive, because in Japanese Buddhist funeral rituals the cremated bones of the deceased are passed between chopsticks. Ask for an extra pair of chopsticks, or let others help themselves from the platter.
- It is permissible to sip from your soup bowl. If you're having noodles in soup, slurping shows how much you are enjoying the meal.
- Eat everything on your plate, right down to the last grain of rice.
- Use the chopstick rest to lay down your chopsticks. Never rest them on your bowl, or worse, stick them straight up in the bowl of rice.
- Tipping is rare in Japan and is considered rude.

CHINA

- Always serve elders first.
- Dishes are all served at the same time, family style. Take food from the platter that is in front of you.
- Always use the serving spoons or communal chopsticks when taking food from the shared plates.
- It is permissible to sip from your soup bowl.
- Do not use your chopsticks to stab or skewer food.
- Don't finish all your food. Leaving a bit of food on your plate shows your host that there was plenty to eat.

THAILAND

- Eat with your spoon, using your fork to help. Use chopsticks only to eat noodles.
- Dishes are served at the same time, family style, and are meant for sharing.
- Never take the last bite from a shared bowl or platter.

INDIA

- Make sure to finish all your food. Any waste of food is considered extremely rude and is disrespectful to your host.

- Eat only with your right hand. Your left is considered unclean for eating, but should be used to drink or pass dishes.
- Wash your hands before and after eating, and make sure your fingernails are clean.
- Never offer someone food from your plate, even just to taste.

MIDDLE EAST

- Wash your hands before the meal.
- Eat only with your right hand. Your left is considered unclean.
- Don't finish all your food. Leaving a bit of food on your plate shows your host that there was plenty to eat.
- Always eat what is offered to you by the host.

ITALY

- Don't ask for extra cheese for your pasta or pizza unless it is offered, and never ask for cheese on your seafood dish.
- Use only a fork to eat pasta. Twisting pasta with a fork and a spoon is appropriate only if you are a

child. If you need to, twirl a small amount of pasta, clockwise, onto your fork using the rim of the plate.

- Never slurp your spaghetti.
- Never cut your pasta with the side of your fork.

SPAIN

- When you order a drink at a bar, the bartender will likely bring you a small tapas plate, on the house. Any tapas that you order beyond that will be charged to you.
- Never stack your empty tapas plates.
- Do not eat and run. The after-meal period is a time to linger and converse.

FRANCE

- Rest your wrists or forearms on the table, keeping your hands visible. This old custom came about when it was important to know that your fellow diners were unarmed!
- Put your bread directly on the table, not at the side of your main plate.

- Do not cut salad with a knife. If the pieces of lettuce are too large to eat comfortably, use a piece of bread in your left hand to hold them in place while you fold them into a smaller bite with your fork.
- Do not ask for salt and pepper if it isn't on the table. It is an insult to the chef, implying that the food was not seasoned properly.
- When eating soup, skim towards you, and sip from the end of the soup spoon. When you are almost finished, tip the bowl towards you to get the last few drops.

- It is considered crass to split the bill; one person should pay it.

NETHERLANDS

- As in France, keep your hands on the table, not on your lap.
- Do not cut salad with a knife. If the pieces of lettuce are too large to eat comfortably, use your knife to fold them.

- Use a fork and knife to eat most foods, including pizza, sandwiches and fruit.
- Do not leave the table during dinner; it is considered rude.

MEXICO

- Do not eat and run. Dining is an opportunity to spend relaxed time with family and friends.
- The proposing of toasts is done by the men.
- As in France, keep your hands on the table, not on your lap.
- Leaving a clean plate is considered rude.
- It is perfectly acceptable to be at least an hour late for a social engagement.

ENGLAND

- Keep your hands below the table in your lap when you are not eating, and keep your elbows off the table.
- When eating family style, offer the dishes to your fellow diners before you help yourself.
- Do not sprinkle salt over your food. Place some salt on the side of your plate and season each bite as you eat.

- Do not cut up more than three bites at a time.
- When eating soup, skim away from you and sip from the side of the soup spoon. When you are almost finished, tip the bowl away from you to get the last few drops.
- When you are done eating, place your unfolded napkin to the left of the place setting.

EATING HEALTHY AWAY FROM HOME

As a business traveller, I used to find myself eating all kinds of foods without a thought. Then all of a sudden I realized I had put on so much weight that my clothes were no longer comfortable. So, I quickly developed some golden rules for eating when travelling. I won't deny myself the occasional indulgence, but these are the guidelines that keep me on track.

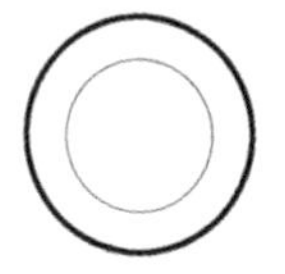

- At a buffet restaurant, I stick to a one dinner plate rule, no matter how good the food is!
- When I'm out, I'll skip the snacks during the cocktail hour.

DID YOU KNOW?

The average cruise passenger gains a pound a day. Self-control is a challenge when tempting and unlimited meals and snacks are available 24/7.

• I refuse to open the mini-bar because it's far too tempting.
• I'm no longer afraid or embarrassed to ask a waiter or room service to have my food steamed, broiled or grilled.
• If I'm having trouble controlling myself I will order two appetizers for my meal rather than one appetizer and one main course. This is all about volume and portion control.
• I always bring along a few packs of protein shakes. This is a good alternative to snacking on doughnuts or warm cookies during mid-morning or mid-afternoon slumps.
• I keep myself well hydrated. Water keeps me feeling full and healthy. For a change, I will sometimes add a couple of drops of flavouring.
• On occasion, if there is something that I really want to have on a trip, I'll plan for it and allow myself the

indulgence without beating myself up. Life is short, so everything in moderation—including moderation!

ETIQUETTE OVERSEAS

Today, the English language is spoken in more parts of the world than ever before. However, that's no excuse for not learning a few key phrases in the local language. Even a feeble attempt makes a good impression and will win you the respect and sympathy of locals.

Handy resources include the traditional phrase books, but you can also find picture guides with hundreds of icons. Just point to the picture to communicate with the locals in any country. If you have a smartphone with a roaming plan, the Google Translate app is free, and it is a lifesaver. I have used this around the world on many trips, and it is an incredible way for travellers to communicate in multiple languages. You can even get a premium version of Google Translate that works without Wi-Fi access!

OVERCOMING LANGUAGE BARRIERS

Below is a chart with a few basic words that everyone should know.

English	French	Italian	Spanish	Mandarin
Hello	Bonjour	Ciao	Hola	Nǐ hǎo
Good evening	Bonsoir	Buonasera	Buenas tardes	Wǎnshàng hǎo
Goodbye	Au revoir	Arrivederci	Adiós	Zàijiàn
Thank you	Merci	Grazie	Gracias	Xiè xiè
Please	S'il vous plaît	Per favore	Por favor	Qǐng
Sorry	Excusez-moi	Mi dispiace	Lo siento	Hěn bàoqiàn
Pardon me	Pardon	Mi scusi	Perdone	Duìbuqǐ
Where is . . .	Où se trouve	Dove si trova	Dónde está	Nǎlǐ
How much?	Combien?	Quanto?	Cuánto?	Duōshǎo?
I would like	Je voudrais	Mi piacerebbe	Me gustaría	Wǒ xiǎng yào
Cheque, please.	La facture, s'il vous plaît.	Il conto, per favore.	La cuenta, por favor.	Qǐng jié zhàng.
Help!	Au secours!	Aiuto!	Auxilio!	Jiù mìng!
Cheers	À votre santé	Saluti	Salud	Gānbēi
Where is the toilet, please?	Où se trou-vent les toi-lettes, s'il vous plaît?	Dov'è il bagno, per favore?	Dónde está el baño, por favor?	Qǐngwèn nǎlǐ yǒu xǐshǒujiān?

CONVERSATION-STARTERS

Starting a conversation with a new acquaintance can sometimes be uncomfortable. When meeting someone for the first time overseas, try not to be too personal. In North America, we tend to be much quicker to ask personal questions. Do your best to avoid asking questions such as: What do you do for a living? Are you married? What kind of car do you drive?

Here are some conversation-starters for you to try with new acquaintances.

- Did you grow up in this country?
- Do you travel often? Where do you like to travel to?
- Are you interested in music? What kinds do you enjoy?
- How do you know our host?
- What is the weather normally like here at this time of the year?
- What is the most beautiful place you have ever travelled to?
- What did you do on your last vacation?

BUTLER'S TIP:

The best way to start a conversation is to ask open-ended questions. Closed-ended questions will yield one-word answers: yes, no, maybe.

SHAKING HANDS

For many of us in North America, shaking hands is a customary form of greeting, but that isn't the case around the world. Be aware of cultural differences and know what is the appropriate greeting before meeting people for the first time.

A few simple rules to remember:

- In business, men and women are considered equal, so either person can initiate a handshake.
- Socially, it is still always safer to let the woman extend her hand first.
- In the Middle East, parts of north Africa and some Asian countries, touching between men and women may not be acceptable.

Here is the best way to execute a polite, professional handshake:

Keep your right hand open.

Connect the joint of your thumb to the thumb joint of the other person, web to web or palm to palm.

Gauge the duration of the handshake by paying close attention to the other person. A good rule of thumb is to pump your hand three times, or for about three seconds.

EXCHANGING BUSINESS CARDS

The new global standard is the Asian way of offering your card. It is easy to learn and respectful towards others as you travel around the globe.

ALWAYS present your business card using both hands, with your thumbs on top of the card. The information should face outward, so the recipient can read the card without needing to turn it.

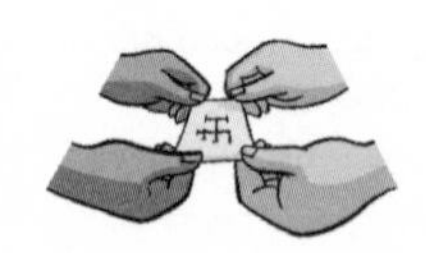

ALWAYS receive your contact's business card using both hands. In some cultures, the business card is considered a representation of the owner and therefore should be treated with respect. Make a point of reviewing the card and commenting on it before putting it away in a card case. Never put it directly into any pocket.

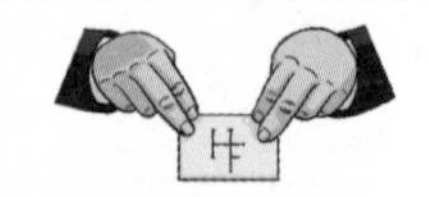

DO NOT present your business card with your fingers at the top of the card. This is inappropriate.

DO NOT present your business card with your hands at the bottom of the card. This is inappropriate.

DO NOT present your business card using just one hand. This may be perceived as arrogant or disrespectful.

GOLDEN RULES OF FILING A COMPLAINT

To complain or not to complain? That is a dilemma most of us face when things go wrong. If you're able to let the incident go and get on with your day, then embrace your positive attitude. But if the issue in question is making you feel uncomfortable or keeping you up at night, deal with it as soon as possible. Here are just a few tips that work well in any situation.

◆ ◆ ◆

Speak up early. If you wait too long to make a complaint, the establishment might not be able to resolve the issue. This certainly is the case with restaurants. If my steak is overcooked, I will politely bring it to the attention of the waiter, who will immediately address it with the kitchen.

◆ ◆ ◆

Remain calm. People are more receptive when you communicate in a clear and reasonable manner. People are more helpful when you are pleasant to deal with.

◆ ◆ ◆

Be prepared. Have all your facts ready, including any pertinent receipts or documents to support your complaint.

GUIDE TO TIPPING

	North America	UK
Restaurant	15–20%	10–15% if no service charge is included / no tipping in pubs
Bellman	$1–$2 per bag	£1–£2 per bag
Housekeeping	$2–$5 per day	£1–£2 per day
Room Service	10–15% if no service charge is included	10–15% if no service charge is included
Concierge	$10–$20 per special request, depending on the service provided	£10–£20 per special request, depending on the service provided
Doorman	$1–$2 for hailing a cab / $1–$2 per bag for carrying luggage	£1–£2 for hailing a cab / £1–£2 per bag for carrying luggage
Taxi	10–15%	10–15%
Hairdresser	15–20%	10%
Tour Guide/ Driver	15–20% of excursion fare	10% of excursion fare
Valet Parking	$2–$5	£1–£2

Europe	China*	Hong Kong
10–15% if no service charge is included	None	10–15% if no service charge is included
€1–€2 per bag	5–10 RMB per bag	HK$10–$20 per bag
€1–€2 per day	10–15 RMB per day	HK$10–$20 per day
10–15% if no service charge is included	None	10–15% if no service charge is included
€10–€20 per special request, depending on the service provided	None	HK$20–$50 per special request, depending on the service provided
€1–€2 / bag	None	HK$10–$20 for hailing a cab / HK$10–$20 per bag for carrying luggage
10%	None	Not expected: round up from the fare, and add extra for heavy suitcases or packages
10%	None	10%
10% of excursion fare	80–150 RMB per day for a private tour guide and half that amount for the driver	HK$100–$150 for a private tour guide, and half that amount for the driver
€1–€2	5–10 RMB	HK$10–$20

(continued)

	Australia	New Zealand
Restaurant	10–15% for exceptional service	10% for exceptional service
Bellman	AU$1–$2 per bag	NZ$1–$2 per bag
Housekeeping	AU$1–$5 per day	NZ$1–$5 per day
Room Service	10% if no service charge is included	10% if no service charge is included
Concierge	AU$10–$20 per special request	NZ$10–$20 per special request
Doorman	AU$1–$2 for hailing a cab / $1–$2 per bag for carrying luggage	NZ$1–$2 for hailing a cab / $1–$2 per bag for carrying luggage
Taxi	Round up from fare, or 10%; add extra for heavy suitcases or packages	Round up from fare to the nearest $1–$2; add extra for heavy suitcases or packages
Hairdresser	10–15%	10–15%
Tour Guide/ Driver	10%	10%
Valet Parking	AU$1–$2	NZ$1–$2

NOTES

** Though tipping is not the custom in China, it has become expected in high-end western-style hotels.*

Similarly, in Japan tipping is not customary, but it is accepted (10-20%) in luxury, full-service traditional inns (ryokans) and at high-end restaurants. Always offer the tip in a small envelope and present it discreetly. Handing out money without an envelope is considered impolite. It is also polite to tip ¥2500-¥5000 per day for a private tour guide and half that amount for the driver.

Middle East	Brazil	South Africa
15–20% in addition to the service charge	10% service charge (generally included)	10–15% if no service charge is included
US$2–$3 per bag	US$2–$3 per bag	R10 per bag
US$2–$5 per day	US$2–$5 per day	R30–50 per day
10–15% if no service charge is included	10–15% if no service charge is included	10–15% if no service charge is included
US$10–$20 per special request	US$10–$20 per special request	R100–200 per special request
US$2–$3 for hailing a cab / US$2 per bag for carrying luggage	US$2–$3 for hailing a cab / US$2 per bag for carrying luggage	R10–20 for hailing a cab / R10 per bag for carrying luggage
Not expected; round up from fare; add extra for heavy suitcases or packages	Round up from fare, or 10%; add extra for heavy suitcases or packages	10%
10%	10%	10%
10%	10%	10%
US$2–$5	US$2–$5	R20–50

The numbers are suggested minimums to use as guidelines when calculating your personal tips.

Tips for housekeeping staff should be left in an obvious place with a note saying "Housekeeping—thank you."

The amount you tip a concierge can depend on the service that is being provided.

A tip for valet parking is offered when the car is returned to you.

In Case of Emergency

An emergency can be a nasty interruption to your trip. Keeping a cool head is the only way to navigate through a crisis, and that takes preparation.

Always keep copies of important travel documents in case something unfortunate happens, and make sure that a trusted person at home also has copies of your documents on hand.

- Passport
- Full itinerary, with contact details and addresses
- Travel and health insurance papers
- Travelling tickets (plane, train, boat, etc.)
- Names of doctors and lawyers

In need of assistance in the event of an emergency, contact your country's embassy or consulate. The contact information is available on your government website. Your hotel concierge might also be able to look up this information for you. Contact your embassy or consulate if any of the following situations should arise:

- Hijacking, hostage-taking, kidnapping
- Large-scale emergency
- Death of a travel companion
- Arrest and detention
- Missing persons
- Child welfare, abduction or custody issues
- Sexual assault
- Forced marriage
- Physical assault
- Serious sickness or injury
- Urgent need of financial assistance
- Stolen passport
- Lost or stolen belongings

LOST PASSPORT

Losing a passport when travelling abroad can be a real hassle. Before your trip, check into the process of replacing a passport so that you will know what to expect.

In a perfect world, you will have a photocopy of your missing passport showing your picture, name, passport number, date of issue and date of expiry. Some people have told me that they carry a digital copy of their passport on their smartphone or computer, but I've spoken with several security experts who advise against it. This puts you at greater risk if your mobile device gets into the wrong hands.

DAILY INCONVENIENCES

Not all emergencies are a matter of life and death. Sometimes, they're a case of a suit jacket losing a button on the eve of an important client meeting, or a coffee stain on a dress in the middle of a formal lunch party. Being equipped to handle these kinds of trials gracefully and efficiently will allow you to enjoy your trip without added stress.

HOW TO SEW ON A BUTTON

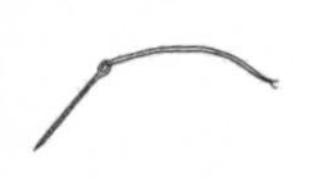

Step #1

Pick a thread to match the button or fabric and cut off about 18-20 inches. Thread the needle and knot the two ends together to have a double strand of thread.

Step #2

Mark the position of button with two straight pins on the right side of the fabric. If the shirt is silk, silk pins must be used.

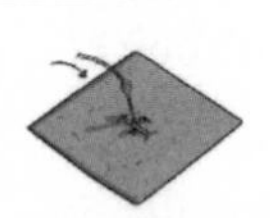

Step #3

Turn the fabric over to the wrong side and insert the needle in a corner close to the intersecting point of the two pins.

Step #4

Turn the fabric so the right side is facing up. Lay the button in position on top of the pins. Pull the needle up through the fabric and one hole of the button.

Step #5

Bring the needle down through another button hole and fabric. Make three or four more stitches to secure this pair of holes.

Step #6

Repeat the process with the second pair of holes.

Step #7

Remove the pins.

Step #8

To make the shank, pull the needle out between the button and fabric. Lift the button so the threads are taut.

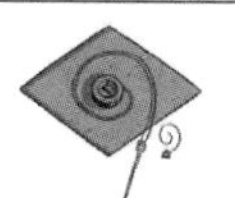

Step #9

Wind the thread tightly several times to make the shank so the button stands up on its own.

Step #10

Pull the needle down through the fabric to the wrong side of the fabric.

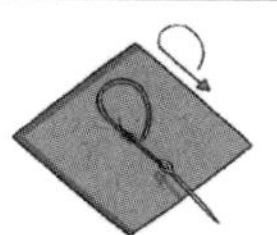

Step #11

Make a partial short stitch under the button stitches and pull the thread through the loop to make a knot.

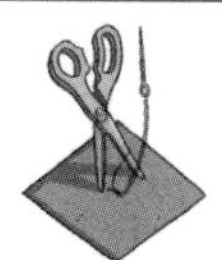

Step #12

Cut the thread as close to the knot as possible.

NOTE: Sewing a two-hole flat button basically follows the same process. However, the two-hole button requires extra attention in sewing it in place. Make sure the holes are sewn in place so they rest in the same direction as the buttonhole.

HOW TO REMOVE A STAIN

If you have room in your luggage, consider packing a compact stain-removal kit. A stain remover pen is a handy quick fix, but, depending on the type of stain, you may need other materials as well. I always travel with a stain remover pen and wipes, a white cotton towel or napkin, pH-neutral dish soap in a small travel-sized bottle, cotton swabs and a plastic knife.

Before using any product on a garment, always test the treatment in an inconspicuous spot to check the fabric for colourfastness, and use the least aggressive method first. And of course, treat a stain as quickly as possible to get the best results.

The following steps work well for most stains, including coffee, salad dressing, butter, gravy, chocolate, lipstick or makeup. Be aware that these are just first-aid treatments that apply to washable garments. If the stain persists, you may need to send the garment to the dry cleaner's once you get home.

1. Place an absorbent white cloth or towel under the stain to prevent it from transferring to other parts of the garment.
2. Blot up the excess, or use the plastic knife or the edge of a credit card (any dull-edged object will do) to remove it.
3. Flush the stained area (back side facing up) with cold water.
4. Working from the outside of the stain inward, gently rub the back of the stain with a drop of pH-neutral dish soap, using your fingertips to loosen it.
5. Rinse thoroughly.
6. If the stain outline is visible, use a cotton swab dampened with rubbing alcohol to dab the stain outline.
7. If the stain persists, treat with a stain remover pen or wipe.

STAYING HEALTHY

Sadly, we have a running joke in our family that I have seen a doctor in every country to which I have travelled. I've had a procedure for an ingrown fingernail in

Shanghai, received medical attention for an ear infection in St. Maarten, for food poisoning in Rio de Janeiro, and for pneumonia in Johannesburg! As distressing as it was to be sick and away from home, I've been very fortunate to have stayed in good hotels with a doctor on call who would come to the hotel for a fee (usually cash).

Before you go on any trip:

- Make sure you have proper health insurance coverage, especially if you are prone to illness when you're away.
- Make sure you have the contact number for the insurance company.
- If you have even the slightest concerns about your health prior to your trip, see your doctor first to get any medications or prescriptions before you leave.

Here are some tips that will help you from getting sick while you're away:

- Check with your doctor or travel clinic for required vaccinations and any other health concerns that pertain to your destination.

- Drink bottled water and stay hydrated.
- Be careful what you eat. Stay away from potential food poisoning hazards (see page 171).
- Wash your hands regularly (see below) to fend off germs.
- Get plenty of rest. If you're jet-lagged and tired, you have a greater chance of getting sick.

HOW TO WASH YOUR HANDS

Wash your hands whenever possible. Your mother probably said this all the time when you were small, and she was right. Note that to do it properly takes at least 15 seconds.

Wet your hands and wrists with warm running water.

Use one or two squirts of liquid soap.

Vigorously rub your hands together to make a soapy lather.

Scrub in between and around fingers.

Scrub the back of each hand with the palm of other hand.

Scrub fingertips of each hand in opposite palm.

Scrub each thumb clasped in opposite hand.

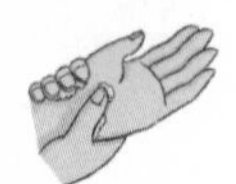

Scrub each wrist clasped in opposite hand.

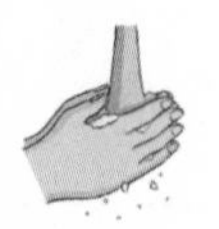

Rinse your hands well under warm running water.

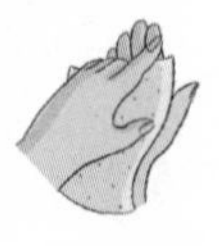

Pat your hands dry with a clean dry towel or paper towel.

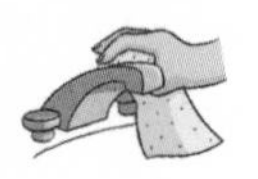

Turn off the water using the same paper towel. This is to avoid cross-contamination from the faucet to your clean hands.

HAND SANITIZER

While never a substitute for proper hand washing with soap and water, hand sanitizers or wipes will work in a pinch when you're out and about. When I'm on the plane, I've even used hand sanitizer on a tissue to wipe down the table, armrests and other surfaces that don't feel clean.

FOOD POISONING AND HOW TO AVOID IT

Contaminated food and drinks cause food-borne illnesses. As tempting as it might be to sample the local specialties served by street vendors and at food markets, you need to be wise.

- Local drinking water is a huge source of illness for travellers. The water may be fine for locals, but the different bacteria and parasites in the local water supply may make visitors sick. Stick to bottled water, even when you're brushing your teeth. And avoid ice cubes, ices and any drinks prepared with local water.
- In locations where water quality is questionable, stay clear of raw fruits and vegetables washed in the contaminated water. Eat only fruits that have peels.

- New and different foods may not always agree with you. Start off by eating smaller portions and eat in moderation.
- Dining in reputable food establishments will be your safest bet. If you can't fight the urge to check out the street food, ask your concierge for advice, or pick the most popular vendor. Before you place your order, assess the food-handling and hygiene practices. Better safe than sorry!
- Make sure the hot foods you eat are hot and fully cooked. Avoid foods that have been sitting out at room temperature, warmed up or reheated. This includes table condiments.

FIRST-AID KIT

Be prepared for the unexpected by packing a first-aid kit. It isn't always easy to find a drugstore when you need one, and if language is a barrier it can sometimes be difficult to find the supplies you need.

Here is a list of items that will be useful in dealing with minor ailments.

ESSENTIAL ITEMS

Pain and fever medication

Nausea and motion sickness medication

Cold and flu medication

Antihistamines

Antibiotic ointment

Hydrocortisone cream

Antiseptic wipes

Bandages

Blister pads

OPTIONAL ITEMS

Saline eye drops

Insect repellent

Sunscreen

Aloe gel for sunburns

Checking Out

BEFORE LEAVING YOUR DESTINATION, ensure that everything that you brought and bought is packed and ready to go home. I've made this mistake twice: once I left behind a necklace that I bought in Rome, and the second time I forgot my medication in the mini-bar of my Madrid hotel. From these two experiences, I've learned to make a checklist the night before of all the items I must not forget to pack, from the bathing suit drying outside in the sun, to medications, chargers, and my passport and other valuables locked in the safe. I love my checklists!

Here's the proper etiquette for checking out of your accommodations.

HOTEL OR ALL-INCLUSIVE RESORT

- Don't leave the room as if a tornado went through it. Yes, there is a housekeeping service, but it is disrespectful to the housekeepers, who now have extra work to clean up your mess.
- Always put your dirty towels in one pile on the bathroom floor.
- You don't need to make the bed, but at least pull the sheets up over the mattress.
- Turn off the lights and television before you leave.
- Try to leave at the designated check-out time so the hotel staff have time to turn over the room for the next guest.
- If your hotel has a self-serve check-out counter, using it will likely streamline your departure.

B&B

- Generally, you don't need to strip your bed, but it's polite to at least pull up the sheets so that the bed doesn't look like you just rolled out of it.
- Try to leave on time so that the host can turn over the room for the next guest.

- Leave a tip for the housekeepers (provided they are not related to the innkeeper).

VACATION RENTAL

- Respect the host's home and put things back where they belong.
- Depending on the host's terms, leave the accommodations in a clean and tidy state.
- Return the thermostat setting to the host's requirements.
- Turn off the lights.
- Leave on time, and let the host know when you left.
- Be honest in your reviews of the accommodation.

FAMILY OR FRIEND'S HOME OR HOME EXCHANGE

- Always ask what you can do. Often family and friends are happy to have you strip the bed and put the dirty towels in the laundry room. If you have time and feel your host will appreciate it, do the laundry, remake the bed and fold the clean towels.
- Tidy up any other areas, such as the bathroom and kitchen.

- Leave a parting gift as a thank-you for their hospitality.
- Send a thank-you note once you are home.

CRUISE OR YACHT

- Generally, you don't need to strip your bed, but it's polite to at least pull up the sheets so that the bed doesn't look like you just rolled out of it.
- Although you tip the ship's crew on your final invoice, remember, it is always appropriate to recognize with a separate gratuity any particular employee who has gone above and beyond for you. Be sure to give this to them discreetly.

RETURNING HOME

“Did you ever notice that the first luggage on the carousel never belongs to anyone?”

ERMA BOMBECK,

Humorist and Newspaper Columnist

Home Sweet Home

AFTER A LONG TRIP, I often return tired and hungry. It can be comforting to know that the basic supplies are already there (for example, milk and bread, soup, a simple supper or fresh fruit). If you have someone house-sitting for you, you could ask them to restock the necessities. Or plan to make a quick stop on the way home—you'll thank yourself for it later.

One of the next things to do is to air out your home. Open a few windows (weather permitting) and let the fresh air circulate. Also, if you adjusted the thermostat before departure, be sure to return it to your normal settings.

CATCHING UP ON EMAIL, VOICEMAIL AND OTHER CORRESPONDENCE

Dealing with a backlog of emails, voicemails and other correspondence can be hard to contemplate when you return from a trip. Here is what I do. First, I listen to my voicemails and make a list of the people I need to call back. Then, I sit with a cup of coffee in hand and return the calls in order of priority. With emails, I'll start to read through them at the airport as I'm waiting for my flight home, or the next day I'll go into the office early and plough through. No matter how you choose to do it, it is courteous to acknowledge everyone, ideally within 24 hours of your return.

SENDING THANK-YOU NOTES

When you return home, thank-you notes should be near the top of your to-do list. People who have hosted you, helped you or entertained you—all need thank-you notes. While a handwritten card is a wonderful thing to receive, an eloquently written thank-you email is perfectly acceptable.

BUTLER'S TIP:

When in doubt about who to send a thank-you note to, send one to everyone. People will often complain they didn't get one, but no one has ever complained about receiving one!

WRITING TRIP REVIEWS

Online reviews are a helpful way to share your experiences with others. A good review is honest, and fair. Never write when you are angry. If your experience was a negative one, take a deep breath, or maybe wait a day or two. If you had a particularly bad experience, it might be best to approach the establishment directly first. How they respond to your feedback may become an important part of your review.

Descriptive reviews are the most helpful. Rather than simply saying you had a good or bad experience, give solid explanations of what you liked or disliked so that your opinion is more meaningful to the reader.

BUTLER'S TIP: GIFTS AND SOUVENIRS

It's important to acknowledge kind-hearted family and friends who looked after your pet or watered your plants, as well as colleagues in the office who covered for you while you were away. A small gift or souvenir from your travels is the perfect token of your appreciation. Ideally, look for items that are locally made, or that reflect the culture of the country you visited.

PEOPLE AND PLACES TO REMEMBER

Acknowledgments

Marcus L.H. Dearn
Marilyn Denis
Julie Fredette
Andrew Gayman
Jody Lee
Julie MacLeod
Rosemary McCallum
Robert McCullough
Dan Mozersky
Diane Muromoto
Judy Muromoto
Madame Neri
Philippe Neri
Peter Papapetrou
Lindsay Paterson
John Robertson
Five Seventeen
Lindsay Vermeulen